PRAISE FOR DOUGLAS McWILLIAMS

"… has a habit in his lectures of dropping in factual nuggets that send a shiver down your spine."
Anthony Hilton, The Independent

THE AUTHOR

DOUGLAS McWILLIAMS is president of Cebr, one of the UK's leading specialist economics consultancies, now one of the most highly respected sources of business advice and research. His career has focussed on making economics relevant to commerce, first as chief economic adviser to the Confederation of British Industry, then as Chief Economist for IBM UK. In 2012 he was chosen from over 300 applicants to become the Gresham Professor of Commerce.

THE FLAT
WHITE ECONOMY

Douglas McWilliams

THE FLAT WHITE ECONOMY

*How the digital economy
is transforming London and
other cities of the future*

DOUGLAS McWILLIAMS

Duckworth Overlook
London and New York

This paper back edition 2016 by
Duckworth Overlook.
First published in in 2015 by
Duckworth Overlook

LONDON
30 Calvin Street, London E1 6NW
T: 020 7490 7300
E: info@duckworth-publishers.co.uk
www.ducknet.co.uk
For bulk and special sales please contact
sales@duckworth-publishers.co.uk,
or write to us at the above address.

NEW YORK
141 Wooster Street
New York, NY 10012
www.overlookpress.com

A catalogue record for this book is available
from the British Library
Cataloguing-in-Publication Data is available
from the Library of Congress

ISBN:
UK: 9780715650653

Typeset by Ray Davies and ketchup

Printed and bound in Great Britain

"*In Silicon Valley we are often too busy bathing in our private sunshine. But the change in London's technology climate has got us checking our weather apps. The weather alerts show that there will be plenty of opportunities to invest in technology companies based in London in the coming years – some of which will become global powerhouses.*"

Sir Michael Moritz, Chairman of Sequoia Capital and author of *Return to the Little Kingdom: Steve Jobs, the creation of Apple and How it Changed the World*

CONTENTS

Contents

FOREWORD

I was prompted to write this book because I was intrigued by the subject matter and my personal connections with it. I was sufficiently excited to break the habit of 40 years and write another book.

Writing books is not something you should do if you want to use your time economically, especially if you had an interesting and exciting day job as I was fortunate to have as Executive Chairman of Cebr, the economics consultancy, when I wrote the first edition.

But this book involves digital technology, which I know about from my years as the Chief Economist for IBM (UK). And the London economy which has been a major interest of mine since my father was Lord Mayor of London and asked me to help him with his speeches on the London economy (for more read the book!). And the epicentre of the Flat White Economy is the London postal district EC1V which has been my own office postal district for the past fifteen years. The subject fascinated me so much that I just had to write about it.

Even as the author I have been astounded by some of the facts I've discovered about London's digital economy. The number of new businesses that have been generated is phenomenal. And the digital economy and its spin-offs are going to dominate not just London but the whole of the UK, let alone other countries. In the book I calculate that about a third of the UK's GDP by 2025 will depend in some way on the digital economy, either upstream or downstream (and nearly 100% will be using it of course!). Since a third of the economy is government (measured by cost, not value) and is not part of the market economy, the calculation implies that half the market economy – that is goods and services produced not financial markets transactions – will depend in some way (other than simply as a user) on the digital economy. It is in some ways already the world's most important sector, and within a decade it will be universally acknowledged as such by pretty well all measures. Already the most recent FT Global Top 500 Companies on 19 June 2015 listed Apple, Microsoft and Google as three of the world's top five companies by market capitalisation, with Apple top, twice as valuable at $724 billion as any other company.[1] Since market capitalisation is a forward-looking indicator (if imperfectly so), this is real world evidence to support my contention.

Most of the themes which I identified for the hardback remain important a year later. And the Flat White Economy has continued to increase in importance. When I made the calculations on 2012 data it was the UK's fifth largest sector outside the public sector with a value added of 7.6% of GDP. Updating to the 2013 data, the Flat White Economy has overtaken the retail sector, the financial sector and the wholesale sector to become the second largest sector in the UK economy outside the public sector, with only construction bigger, and now accounts for 8.7% of GDP directly. This shows how important the sector is to the UK.

But although the past performance has been spectacular, a little of the enthusiasm for the future has waned.
Rising property prices have squeezed out some businesses
and slowed the pace of growth of the industry in East
London while creating increasing resentment towards the
so-called 'hipsters'. The growing focus on limiting immigration is slowing the pace at which skills move to London
and as a result skill shortages are becoming more apparent. And the sector has still to create its first Google or
Amazon although there are allegedly 14 unicorns – companies with a market value of over $1 billion. More than
half the tech unicorns in Europe are based in the UK and
most of these are London based .

There is an interesting question about why the found-
ers of UK tech companies tend to bail out when their
companies are worth around £50 million or so – which
is normally enough to set them up for the rest of their
lives. Do they lack ambition? Or is it perhaps that they are
too civilised to think that growing a business is an end in
itself. I don't know the answer to this. But what I do know
is that if you stop driving your business to grow when it
has reached the point where it has generated sufficient
wealth to provide for your own needs, you are making less
of a contribution to your society, your community and
your employees than you would make if you kept pushing
for further success.

But the good news is that the Flat White Economy is
still driving economic growth in London and in the UK.
Indeed since the first edition was published the growth
has started to spread – new clusters are spreading in
South London and elsewhere in the UK. It is not too

much to say that South London is the new East London.

At the same the emerging sectors in the West Midlands, Manchester, on the M4 corridor and in Brighton which had not been so prominent a year ago have started to flex their wings and I have updated the section on the Flat White Economy in the rest of the UK to take account of this.

The Flat White Economy looks to have grown by an additional 8% in 2015 and to have continued to drive about 40% of the UK's GDP growth for the year. At this rate the sector looks to be in good shape to realise the prediction in the hardback edition that it will drive a third of the UK economy by 2025.

As in the first edition, I would like to thank my publishers Duckworth and my wonderful agent Diane Banks. And also my colleagues especially my friend and chief executive Graham Brough who not only introduced me to Diane Banks but who also has sheltered me from the work that I really ought to have been doing while diverted by writing this book.

But most of all, I would again like to thank my wife Ianthe. She has helped me cope in myriad ways and left me eternally grateful. She wholly deserves the dedication of this updated edition to her.

Douglas McWilliams
London
January 2016

PROLOGUE
Introducing the Flat White Economy

I am writing this from Cebr's main office in Bath Street, London. My company, the Centre for Economics and Business Research has been based here since 6 June 2000. We are in the heart of the postcode EC1V.[1] This small area has 3,228 businesses per square kilometer. This is an astonishing density of businesses - the hightest in the UK.

This book tells the story of what has made such a high concentration of businesses happen and what it means for all of us. And, crucially, how it might change things when it is copied elsewhere. This isn't just a story about a part of East London. It's not just a story about London either. It is a story that will have resonance in urban areas in the rest of the UK and throughout the world.

I first became aware of what was happening because, as part of my job, I am contracted to provide the forecasts for the UK's Passenger Demand Forecasting Council. The PDFC is a consortium of all the UK's key rail transport providers, including the Department for Transport and Transport for London. They get the best economists in the UK to forecast demand for rail services for them: to try to anticipate passenger numbers, in other words. Cebr has been lucky enough to hold this contract since 2011.

When there is an issue about rail demand in a particular part of a city, we will work with the relevant experts to find out what is happening. What we discovered in 2012 was that, in the Old Street Roundabout area of east London, demand seemed to be rising by huge amounts – more than double digit percentage growth.

Normally, when you get surprising data like this, it turns out that there is a simple explanation – a change in the method of recording the data, for instance. Big jumps in economic activity on this scale just don't occur in real life very often. But when we investigated this unexpected growth in passengers through Old Street, we discovered that we hadn't got a recording error after all. The economic activity was for real. It was a genuine massive increase in the number of people working in the area that had generated the increase in tube and rail journeys.

We then looked into what sort of people were driving this growth. Fortunately that was relatively easy, since our office was right in the middle of the area. We looked at what types of people were coming in. They seemed to work in a range of occupations: from advanced techie geeks and marketing people to creative types. What they had in common was something one of the Cebr economists, Rob Harbron, had noticed – a tendency to buy a lot of coffee, and a particular type of white coffee at that. So, since we couldn't find a single industry grouping to name the sector, we decided to called it 'The Flat White Economy'. Within six months the influential economics news organisation Bloomberg had given a conference on the 'Flat White Economy' (FWE). The name had stuck.

There are three ingredients to the Flat White Economy:

technology, demand and skills. Without the technology, there would never have been the business demand that underpins the companies that have started up (we'll get to the skills part in a minute). In the UK, we have been pioneers in taking up new technologies. The UK has been the Western economy that has taken up online retailing fastest. According to the Centre for Retail Research, online purchases accounted for 15.2% of all purchases in the UK in 2015, compared with 12.7% in the US and a European average of 8.4%. Even Germany, by far the most advanced continental European economy in this area, only bought 11.6% of its purchases online.

The UK's growth in online spending has been matched by its growth in online advertising. Again, the UK is a leading economy for this (although spending on advertising in Australia, the US, Canada and Scandinavia is higher per head of population). Online advertising is a critical ingredient in the Flat White Economy because many of the jobs from this economy come from various forms of digital marketing. The high online retail spending and digital advertising spending explains why the Flat White Economy has come to the UK. But not why it has come to East London. Here the third vital ingredient, skills, comes in. London is a Mecca for young people from all over the globe – but especially Europe. It is not cheap – accommodation in particular is very expensive in London compared with almost anywhere else in the world. But East London is the cheapest part of the central areas in London. And the backpacker generation of young people – those that could pack pretty much all their belongings into one backpack – can mitigate the costs of London's expensive rents by drastically economising on space. Youth unemployment rates of over 50% in much of Southern Europe, along with stifling employment regulations that make employers reluctant to hire over much of the Continent, have made London an attractive place to move to for work. EU citizens automatically have the right to work in London. From elsewhere it is more complicated – but, even so, people arrive. And it's not just from abroad: a third of all young people who move location within the UK move to London[2.]

This critical mass of young people means that they have more fun. This critical mass supports a wide diversity of pubs, clubs, bars and restaurants. And with so many young people around, and so many places to enjoy yourself, the statistical opportunity of meeting a potential partner or friend is much higher than elsewhere. The mix of different people with different backgrounds has also created an explosion of creativity. It is this creativity that has been crucial to the development of the Flat White Economy. Skills, talent and a lively labour supply of young creative people – that's why the Flat White Economy is in London. And why EC1V? Why the Old Street area? Well, it is the nearest point of what might loosely be called Central London that's within reach of the less expensive housing in East and North London. That is why Old Street Roundabout has become popularly known as 'Silicon Roundabout.

The scale of this new economy is such that it has affected – directly or indirectly – the whole UK economy. Roughly one third of the UK economy (31.9%) is now in business and financial services: this contributed 42% of the total GDP growth between 2012 and Q2 of 2015. Obviously not all of this growth is from the Flat White Economy. But it is the driving force behind this growth. And it appears probable from my analysis for this book that even the rapid growth officially recorded understates the Flat White Economy's contribution.

The *People's Daily* newspaper – the mouthpiece of the Chinese government – has described the UK as an "old, declining empire", one which resorts to "eccentric acts" to hide its embarrassment over its declining power.[3] But the latest data shows that the UK is one of the faster-growing economies in the Western world. If we compare like with like, we would compare London with Hong Kong, the Chinese city at an equivalent phase of development to the UK: the latest data shows that Hong Kong is growing only at an annual rate of 2.8%[4] while London is growing at about 4%. So the part of China and the UK that can be compared show London, driven by the Flat White Economy, growing one and a half times as fast as Hong Kong. Not bad for an old declining empire.

CHAPTER 1

How the Flat White Economy saved London

I became interested in the London economy by accident. An accident of personal history, that is. My father, Sir Francis McWilliams became, through a fascinating combination of circumstances, Lord Mayor of London for the year 1992/3 described in his autobiography *Pray Silence for Jock Whittington* (2002). At the time, I was the Chief Economic Adviser to the Confederation of British Industry (CBI) and one of the UK's better-known economists. He asked me if I could prepare a basic speech for him on the London economy. Which I duly set about. But what I had expected to be a simple task turned into a nightmare.

I discovered that there was actually very little good economic data available. Neither the post of Mayor of London nor the Greater London Authority existed back then. There was no single body with responsibility for the London economy – let alone for keeping statistics on it. Insofar as it was required, the Lord Mayor of the City of London filled in the role as the representative figure for London as a whole – it was actually my father who submitted London's initial Olympic bid. But my father didn't keep any statistics on the London economy. The research that I thought would take up a spare afternoon became a quest using up most of my waking hours for the better part of a month. But by the end of that month, not only did I have as good an understanding of the London economy as anyone, but I had also estimated many of the basic statistics about London's economy that underpin our knowledge today. Of course, these statistics have long since been updated and improved upon by people with much more statistical talent than I have. But it fell to me to make the first basic calculations. Ever since, I've had a bit of a reputation for "knowing about London" – it works as both a compliment and a putdown, since it enables someone who isn't a fan to dismiss me as "just a regional economist". Actually, I don't mind being called a regional economist. In economic terms, the nation state – the traditional unit for economic analysis – is being broken down because of the impact of globalisation. Globalisation impacts upon different parts of national economies so disproportionately that it increasingly leads to substantially wider gaps between regions in the same country. To ignore these gaps is to miss a lot of what is going on.

The London of the early 1990s was a fascinating economy. It was then in the throes of the 1990s recession, which followed Nigel Lawson's boom. One consequence of the 1990s recession was the collapse in both the commercial and residential property markets as interest rates surged. Because of the historic importance of property in the London economy,[1] it took nearly a decade to work off the excess supply of commercial property in central London and re-establish growth in that sector. But the dominant factor in the economic history of London in this period was the rise of financial services following the "Big Bang" – the deregulation of financial markets in October 1986.

Oddly, I was there right at the beginning of the Big Bang. In a rare invitation to 10 Downing Street in the mid-1980s I had my longest ever conversation with Margaret Thatcher, to which my contribution was only three words – "Yes, Prime Minister". She then left me to cuddle up on a sofa with Sir Nicholas Goodison. Sir Nicholas was then the Chairman of the Stock Exchange, hence the critical figure in the bonfire of regulations that created the Big Bang for London's financial service industry, that explosion of financial service activity in London. Sir Nicholas was a very straight-laced figure who seemed to shrink further and further into the sofa as the Prime Minister flirted with him. Anyway, she had her wicked way with him – he acceded to her requests and the rest is history. I was amused that someone as powerful as the Prime Minister was prepared to use feminine wiles to get what she wanted.

It had become clear to me while analysing the statistics that, since the Big Bang, London underwent an economic growth to rival the mega-growth of Far East economies. I called it the "Tiger Economy on the Thames". London's growth continued until 2007; indeed, for the entire period from 1997–2007, London's economy grew at an annual rate of 6.2% in cash terms. With inflation averaging perhaps 2% over the period this meant real GDP growth grew at an annual rate of over 4% – fast enough to rival Far East economies. This growth was underpinned by the City of London: the number of "City jobs" rose from a plateau of 170,000 in the mid-1980s to a peak of 230,000 in 1989; it collapsed to 190,000 in 1993; but it had recovered to over

350,000 by 2007.[2] As the number of jobs increased, so did the money on the payslips. City bonuses grew from about £1 billion in 1990 to a peak of £14 billion in 2008, before collapsing to £2 billion in 2011. The number of people receiving million-pound bonuses in the City rose from 30 in 2000 to 1,500 in 2007.

From Big Bang to house price boom

But the spin-off from the boom in the City affected the whole of London. People who had never previously had much money spent it wildly. About half of City bonuses went into the property market in some way and London house prices rose wildly to reflect the fact.[3] By 2013, London house prices had risen seven-and-a-half times in 30 years. And they jumped most spectacularly – by four times – between 1995 and 2007. Meanwhile young people with what appeared to be far more money than sense moved seamlessly from the high-adrenaline working world of the City trading floor to the high-adrenaline playgrounds of the champagne bar, the lapdancing club and the Ferrari showroom. The lifestyle trickled down from the traders to the people who serviced their requirements. Stories abounded of supporters of London football clubs waving their wage packets at visiting Liverpool supporters, chanting "Bet you're on the dole!". Shops, bars, restaurants and other service providers in London benefitted financially but suffered morally from the mega-bonus culture that infiltrated London in those days.

People think of the heyday of the "loadsamoney" culture being the 1980s. But the financial merry-go-round in the city didn't actually come to a shuddering halt until the great financial crisis of 2007/08. Bonuses continued to rise and London's economy, driven by financial services, grew much faster than the rest of the UK. Culturally, the noise of "spend, spend, spend" grew quieter – the wealth had started to age and a degree of discretion had emerged – but London's disproportionate wealth continued to accumulate. There are plenty of statistics that show that spending on non-essentials had continued and also that wealth had been moving into property. And London property prices reflected the fact with house prices up by a third since 1983 compared with the national average. By 2014 the average price of a house in London was £330,000 compared to a national average of £182,000.[4] By 2007, someone who had bought a house in Central London in 1990 for £600,000 might well have found that it was worth £4 million. Arguably one of the easiest and least meritocratic ways to get rich then; though, in fairness, those benefiting from the property-price inflation would in all likelihood have had to scrimp and save to pay huge mortgages at times of double-digit rates of interest to hang onto their houses.

The financial crisis of 2007, followed by the banking crisis of 2008, did no one any favours. But the silver lining in the cloud was the possibility of a rebalancing of the British economy: away from financial services, consumption and an overreliance on the property market towards manufacturing industry, investment and exports; and away from London and the South towards the Midlands and the North. For one year, 2009, it happened. In that year London's GDP declined by 6% compared with a fall of only 4.3% for the UK as a whole, while London house

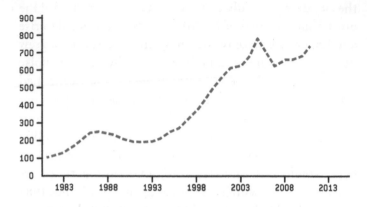

Figure 1.1: House prices in London 1983–2013. Source: Lloyds Regional House Price Data, 2014

prices declined by slightly more – 19%, as opposed to 17% for the UK as a whole. But even the evidence of that year should have told us that a big rebalancing was unlikely to occur. London's GDP and house prices had only declined by a small amount more than the rest of the country's in the face of an economic crisis that had targeted London's twin specialisations, finance and property, with a laserbeam-like intensity. This should have made it clear to everyone that London's resilience was much greater than had been expected. My explanation at the time was that the hits to the financial system were probably not accurately being measured and that the loss of financial activity was badly accounted for, at least in terms of its timing. I also thought that some of the damage to the financial system from the financial collapse would only fully show itself by slowing down the recovery rather than by adding to the immediate impact of the recession. Both conclusions were probably valid, but they failed to explain why London should since have recovered so much faster than the rest of the country.

Recovery: a new economy?

The recovery in London started at the end of 2009. Employment had fallen by a sharp 1.7% that year, but it grew by 0.5% in 2010, with similar growth in 2011. By 2012 employment growth in London was really on a roll, with a massive 2.3% growth. In 2013 employment growth in London was even more spectacular – at 4.4% and in 2014 even more still at an astonishing 5.7%.[5] I've used employment as an indicator because (as I will explain later) there is some doubt about the GDP/GVA[6] data. But the GVA (Gross Value Added) data has also confirmed the picture of London growing noticeably faster than the rest of the UK. The other item of economic data which graphically illustrates regional differences is house prices. By 2012 house prices in London were more than 11% higher than their previous peak in 2007. In the South East of England they had just surpassed their previous peak and were 1% above. For most of the rest of the UK, house prices remained between 5 and 10% lower than in 2007. Evidence that London was leading the UK economic recovery at a time when regional rebalancing had been expected started to emerge at the same time as the dramatic transport data in Old Street described in this book's Prologue. We started to put two and two together and – unusually for economics, where things don't normally fit so neatly – made four.

The spectacular growth in people travelling to work in the area around Shoreditch and South Islington was reflected in a growth in office rents. For a time in autumn 2013, office rents near Old Street were higher than near St Pauls in the heart of the City of London. This was a far cry from the world (as recently as the mid-1990s) when the Corporation of London started to try to provide development assistance to support economic growth on the so-called, depressed 'City Fringes'. Meanwhile, another clue to what was going on came from the sectoral data on jobs growth in London. The growth between 2010 and 2014 in the number of jobs in the "professional, scientific and technical" sector, the largest single component of the Flat White Economy, was a stunning 79,700.[7] And a further 50,000 jobs were created in the administrative and support services sector (which contains most of the rest of the jobs encompassing the Flat White Economy, though also a fair number of other jobs outside the sector).

The heart of this new economy is digital retailing and digital marketing. These jobs are in the UK because the UK is a world leader in both, excelling particularly in the former. But the other side of the coin is creativity. The type of people that have caused London to become an epicentre of digital creativity are young and come from all around the world. London is an essentially migrant-based labour market; four out of every ten employees in London were born outside the UK. Of these, about three-quarters were born outside the European Economic Area (EEA), the remaining quarter from the other European countries in the EEA. Of course, it is true that many (around half) of those working in London but born outside the UK work in low-skilled jobs. But the other half work in skilled jobs and are often particularly highly qualified. Even among migrant workers in low-skill employment, a surprising large number have academic qualifications well in excess of those required for their jobs.

Although the pace of non-EEA immigration slowed after the election of a Conservative-led coalition which cut back on those parts of immigration subject to UK control, the growing economic recession in Europe associated with the problems of the euro led to a massive rise in immigration to the UK from within the EU (especially from Southern Europe). This was especially true since labour market restrictions that are conventional in continental Europe tend, in a recession, to protect existing jobs at the expense of new entrants to the labour force – which causes the employment burden of recession to be felt much more intensely in youth unemployment. The solution for many unemployed youths in continental Europe was to come to London. English is generally the second language for most of Europe so there isn't much of a language barrier. Because of the more flexible (read greater hiring and firing) nature of the UK's labour market, there are many more jobs available there, and especially in London. The city also has a well-deserved reputation as being the party capital of Europe (excluding holiday resorts like Ibiza) – as early as the 1990s once the Eurostar had got into full operation, as many as 50,000 French youths visited London every weekend to go to the clubs, bars and pubs and enjoy the 24 hour weekend city.

The Flat Whiters

Whereas in an earlier era, the more skilled of London's migrants looked for work in the City of London in financial services, post-financial crisis these jobs were few and far between. But one sector was recruiting – the digital economy. And young people from all around the world found that their day-to-day life skills in negotiating the apps on their variety of digital devices made them attractive propositions to the new startups that were servicing the UK's fast growing online retail and online advertising economy. When these skilled and energetic young people from all around the world started to work together, another virtue of the migrant economy became apparent. Not only did migrants provide skills but they also stimulated creativity. People with different backgrounds and ways of thinking spurred each other on to produce ideas.

Creativity is the backbone of the digital economy. Consider MindCandy, the company that created the digital cartoon characters Moshi Monsters, founded by digital entrepreneur Michael Acton Smith. They started in South London in 2004. By 2008 their staff numbers had risen to 18. In 2011 they moved to the Tea Building on Shoreditch High Street (the heart of the Flat White Economy) and staff numbers had risen to 80. By 2013 they had moved their office again to Bonhill Street (just off the Old Street Roundabout) and staff numbers were up to 180. The people at the heart of the Flat White Economy are very different from those who worked in the burgeoning City of London a few years earlier. The Ferraris, champagne and mansions in Kensington and Notting Hill have been replaced by Oyster cards, bicycles and shared flats in Hackney.

The economics of a Flat White Economy startup are less obviously promising than those of businesses in the City. City businesses have always charged premium prices and paid premium wages and bonuses. For a Flat White Economy startup, the business model is much more labour intensive. You need lots of employees, rather than a few very highly paid employees. And the profits of the FWE startup are much more skewed in a "winner-takes-all" direction. Many businesses have relatively few sales initially, during their research phase, and only generate an income later on. This income might be average but, for the few who hit the jackpot, the sales revenues and profits can become enormous. So the FWE startup typically pays badly – just enough for the employee to subsist in a backpacker's lifestyle in London's expensive property market. Shares, options and bonuses are promised when the business takes off. Often rather shadowy financial backers manage to wriggle out of such commitments although, to be fair, take-off is often so late that many of the original staff have long left before payments would be triggered.

So pay is typically low. And the lifestyle reflects the pay. Flat Whiters typically live in North East or East London and share flats (and even rooms). Compared with other Londoners they economise massively on property costs by living out of a suitcase. Their minor extravagances tend to be high-tech IT kit – the latest phone, tablet or laptop and, of course, the latest apps. Many cycle to work wearing high-tech cycling gear. But they do spend money on partying, bars and clubs. And they tend to have sophisticated tastes when eating out. Whitecross Street Market – just off Old Street – which as recently as 2010 mainly consisted of traders selling tat or items that had fallen off the back of a lorry, is now London's best lunchtime food market, with food carts selling a wide range of tasty delicacies for lunch (à la Portland, Oregon).

But while Flat Whiters do not quaff champagne and are unlikely to even own a car (let alone a Ferrari or a Porsche), the Flat White Economy operates on so big a scale that it affects the UK macroeconomy. In 2014 the UK emerged as one of the fastest-growing economies in the Western world, growing roughly twice as fast as the Western average according to Cebr estimates and forecasts and substantially faster than the Eurozone, where growth is close to a standstill. About a third of this growth surplus of the UK reflects the success of its digital economy. Moreover, my calculations suggest that even these numbers actually underplay the real contribution of the UK's digital economy, which is more than twice the size officially measured. The failure accurately to measure the growth in the economy is one reason for the UK's poor measured productivity performance. In reality the UK, even on the official figures, is steaming ahead rapidly compared with most other Western economies. With the digital economy more accurately measured, its outperformance is even more spectacular. And as the oriental economies slow down, the UK's revival is even more outstanding.

But as we move into 2016 it is important to remember the challenges that will have to be met to sustain the impact of the Flat White Economy. I warned in the first edition of the book that there were three weak points that might undermine the growth of the Flat White Economy – property costs for both commercial property and for housing, lack of transport capacity and potential restrictions on immigration.

The evidence of the past year is that all three are starting to have an impact. A report by Cebr and London First in October 2015 has shown that lack of housing is costing the London economy at least £1 billion per annum[8]. In September 2015 the accountancy firm Deloittes reported that 5% of its 2014 intake of graduate accountants had to share bedrooms because of the cost of housing, while in January 2015 the Guardian reported a 74% increase in the number of people looking to share a bedroom[9]. It appears that these people were sharing bedrooms for economic reasons, not for other reasons.

Meanwhile road congestion in East London has reached the point where the average tailback for the Blackwall Tunnel in October 2015 was 2 miles[10] while congestion in Old Street station at rush hours is such that potential passengers normally have to wait for at least two trains before they can squeeze themselves in.

Although rules affecting migration into the UK have not changed since the first edition of the book was written, much discussion of the UK's links with the EU has centred on whether the UK can halt migration from other EU member states. With a referendum on the UK's membership of the EU planned for before the end of 2017, this must act as a worrying signal to those considering investing in the UK's Flat White Economy, especially in London which depends on immigration for its skills.

It is likely that the growth of the Flat White Economy in East London is now past its peak for these reasons. But the concept remains alive. New areas are inventing Flat White Economies of their own – South London is starting to become the new East London. And elsewhere in the UK the economy is starting to take off – Birmingham and Manchester are particularly prominent.

What is clear now is that whereas in its initial stages the growth could happen with little involvement from the authorities, to sustain future growth planning and transport in London will need to work hard to accommodate the sector. Without this, there is a risk that not only will the growth subside but that the sector itself might decamp elsewhere.

CHAPTER 2

Why the Flat White Economy
came to London

There are three factors that brought the FWE to London: the timing and take-up of digital technologies; the speed with which the UK has taken to online retailing and marketing; and the availability of a labour force with a high level of available skills and a high level of creativity. London has the most creative labour force in the world; creativity is crucial to the digital economy.

When I was Chief Economist for IBM (UK) in the mid-1980s I was assured by techies in the company that the digital economy would take off in the next ten years. I was sceptical because I expected that something that would be happening within the next ten years would surely already be starting to show some lead indicators already.

In the event, the digital economy took off about ten years later than the techies had predicted. It is only now, about a quarter of a century later on, that its true worth in sales and advertising is starting to come to fruition. One of the reasons why the digital economy has emerged more slowly than had been expected by the techies is that for take-off to occur for a technological product or service it not only has to be technically feasible but also economically and commercially viable as well. And the economics underpinning the digital economy are actually not all that simple.

One of the most crucial economic factors in the success of the digital economy is massive economies of scale – what I call 'supereconomies of scale'. The concept of economies of scale developed at the beginning of the industrial era. The concept is mentioned in Adam Smith's *Wealth of Nations*, where he points to economies of scale resulting from the division of labour. The concept has typically been associated with the intensive use of machinery, wherein long production runs of a commodity reduce the unit cost of producing that commodity.

But in the information era, wherein the product is not a physical one, the achievable economies of scale are much greater. The underlying economics of information were described in Nobel Prize-winner George Stigler's seminal article in 1961.[1] Typically an item of information is expensive to research and develop but can be replicated through dissemination at virtually zero cost. To give one example, it is believed that the 1990s version of Microsoft Windows (3.0) cost about $550 million to develop. But to create an extra licence cost only a small administration, packaging and copying fee of a dollar or two at most. At the time an operating system like Windows typically sold for about $70. So had Bill Gates only sold a million licences, he would have been sitting on a big loss – of nearly $500 million. But he didn't, of course: he sold about 10 million licences in the two years before Windows 3.1 was released and ended up with a substantial profit that changed the

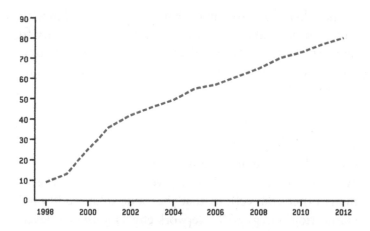

Figure 2.1: UK Households with Internet Access (%). Source: www.ons.gov.uk

future of his company. By the time of Windows 7 and Windows 8 the typical first two years' sales of a Windows version were about 200 million.

The point here is that for the digital economy, massive sales are necessary to amortise research costs. This means that companies tend to delay the investment until they are certain of two factors working in their favour: firstly, that sales will be on a sufficient scale for them to make a profit; and secondly, that there will not be so much competition that they will be forced into a price war. This takes time (although sometimes the desire to keep ahead of competition will encourage companies to bring forward innovations). In my experience, supereconomies of scale have tended to mean that large changes tend to be delayed while incremental improvements tend to be accelerated to keep ahead of competition.

The other economic factor that affected the timing with which the digital economy bore fruit is its sensitivity to network effects. Network effects are a phenomenon common to communications systems. Essentially, the value of a network to an individual participant in that network increases with the number of participants in it. So a single telephone is of no use on its own – it only gains some value when there is a second phone. And as the value of each phone increases, the more phones tend to become available to contact on the network, until there are so many other connections that the value of an additional connection is negligible. Network effects were first developed as a concept by the President of Bell Telephones in making his case for a monopoly in 1908, but the ideas were developed and refined in the 1980s and 1990s. Robert Metcalfe, one of the co-inventors of the Ethernet, was the progenitor of Metcalfe's Law – that the value of a communications network varied with the square of the number of connections in the network. This idea was vigorously promoted by the economic guru George Gilder[2] during the 1990s.

Where there are network effects, investment typically doesn't take place until there is a critical mass of potential users. Network effects tend to cause investment to be held back in a similar fashion to supereconomies of scale, although in the case of the latter, the tendency to delay investment is moderated by the possibility of gaining first-mover advantage. It is the combination of supereconomies of scale and network effects that meant that the economic exploitation of the digital technologies took place on a different, tardier timetable than that which

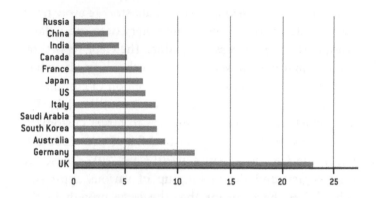

Figure 2.2: Internet shopping as % of total retail in 2016 according to BCG. Source: www.bcg. com (The Connected World: The Internet Economy in the G-20)

had been predicted by those who only understood the technological issues.[3]

Online retail and marketing

Internet usage in most Western economies really took off during the first decade of the 21st century. In the UK, for example, roughly 50% of adults over 16 (and possibly a higher proportion of those below 16) used the internet every day, while the proportion who never used the internet fell down to only a quarter. But the intensity and usage of the internet for commercial purposes only started to grow massively towards the end of the decade. In part this was because of the coming together of a range of digital technologies. Cloud computing, for example, has taken off in the past five years, moving data storage away from the hard drives of personal computers onto mass databases, hosted and accessible online. The Cebr report[4] on cloud computing showed that this technology alone had the potential to generate over €750 billion of cumulative economic benefits and an additional 2.4 million jobs. Big data, which enables online retailers to target customers, was forecast by Cebr[5] to yield benefits of over £200 billion in the UK alone between 2012 and 2017. These big technologies, and the picking-up of various app-based technologies, have meant that the large growth in the scale of commercial activity on the internet in the Western world, and particularly in the UK, has really been in the period since around 2008 till now. The growth is likely to continue, though probably at a more sedate pace in the coming years.

In 2012 the OECD estimated that the proportion of British people shopping online was twice the OECD average.[6] According to the Centre for Retail Research, in 2014, 67% of British people shopped online on a regular basis.[7] The Boston Consulting Group even predicted that by 2016 the UK will have nearly double the proportion of internet retailing that most Western economies will have.

No one has a particularly good explanation for the high proportion of British online spending. The UK has some of the most congested road traffic systems in the world, which must discourage shopping in person in some parts of the UK. And British people are, tradition decrees, temperamentally diffident about face-to-face dealings, which may also encourage online spending over and above purchasing in person. Throughout the world, consumers are concerned about internet security but, by and large, people in the UK are comfortable with the existing systems for using credit cards online – many in other countries are more cautious. Whatever the reason, it's the high level of UK spending online that gave rise to the UK's big potential for online marketing.

The UK has traditionally been strong in marketing services. Its advertising industry has won more awards for creativity than that of any other country in the past ten years according to 'Advertising Pays',[8] a report by Deloitte on the economic impact of the UK advertising industry. This report goes on to argue that, adjusted for GDP, the UK advertising industry "leads the world". The evidence for this can be found in the estimated £2 billion of exports of advertising services from the UK. The UK is also strong by international standards in other areas of marketing. For example, one third of the value added from its market research companies was earned overseas – which would imply a substantial level of exports from the sector.[9] The combination of all these factors meant that the UK was the most likely location for the growth in the digital marketing economy. And with nearly 40% of the UK's professional service activity based in London,[10] London was always likely to be a probable host for such activities.

One of the critical ingredients in marketing is creativity. This theory is best established in advertising, wherein creativity is highly valued and reflected and encouraged by awards. One Harvard Business School study, by Werner Reinartz and Peter Saffert,[11] examined 437 German TV advertising campaigns between January 2005 and October 2010. Their findings "confirm the conventional wisdom that creativity matters" and that "overall, the more creative campaigns were more effective". An Australian study found that creative advertising was four times as effective as other kinds, especially looking beyond a six-month horizon.[12]

Migration and creativity

The importance of creativity in advertising appears to be growing. An historical study of creativity in advertising shows how the role of creativity and innovation has increased since the 1960s.[13] And, given the nature of the medium, it would seem probable that a world like the present one, in which the shape of advertising is adjusting from traditional media to digital media, would place an even bigger weight on creativity than earlier periods. Meanwhile, if anyone were in any doubt as to the links between migration and creativity, there is a preponderance of articles out there. "All immigrants are artists" is a fascinating assertion by the Haitian-American author Edwidge Danticat[14]: perhaps it overstates the case, but it certainly draws our attention to the powerful link between migration and artistry. London has an entire museum, the Ben Uri Museum, dedicated to the theeme Art, Identity and Migration. One of the major themes of study by the Research Council for the Arts and Humanities in the UK has been the link between migration and art under the 'Diasporas, Migration and Identities' programme.[15] Although creativity in the arts is not quite the same thing as creativity in advertising and marketing, it is noteworthy that psychological studies are prepared to bundle them together. One US study by the National Institute of Economic and Social Research (NIESR), carried out for the Migration Advisory Committee, concluded that: "skilled migrants often make positive contributions to innovation and productivity performance. Therefore, whenever restrictions on immigration are contemplated, it is important to assess the potential economic consequences of such restrictions".[16] A study conducted in 2002 in Silicon Valley in the US discovered that Chinese and Indian immigrant engineers had been disproportionately responsible for innovation and entrepreneurship in the area.[17]

The role of migrants and immigrants in London's economy

The London economy has traditionally catered for, and been boosted by, migrants – whether from elsewhere in the UK or from other countries. But the growth in international migration since the second half of the 20th century has turned a trickle into a flood. Between 1986 and 2006 the number of London residents who were born outside the UK nearly doubled – from 1.17 million to 2.23 million – and the proportion of London's residents born overseas rose from 17.6% to 30.5%.[18] The number of migrants grew substantially following the accession of Poland, among others, to the EU in May 2004, subsequently followed by several other Eastern European economies.

The onset of the financial crisis and the imposition of stricter immigration controls on non-EEA migrants slowed the movement of international migrants to London but the emerging economic crisis associated with the euro in many parts of continental Europe has caused a new wave of migrants, especially from Southern Europe, post-2010. By the time of the 2011 Census, 38% of London's workforce had been born outside the UK compared with 15% for England and Wales as a whole.[19] It is not only migration of an international kind that London benefits from. The Centre for Cities claims that a third of people in their twenties who move between cities head for London.[20] Typically, London boroughs such as Camden, Islington, Wandsworth and Lambeth find that just under a tenth of the resident population in any given year has moved to the borough within the past twelve months.[21]

Migrants from abroad are typically more skilled than the local population. Across the UK as a whole, 46% of new and recent immigrants in 2008 had degree-level qualifications or above compared with 18% of the UK born labour force. And migrants to London were more than one-and-a-half times as likely to possess a degree or higher qualification than those migrating to the rest of England and Wales.[22] One of the consequences is that London's labour force is very much more highly skilled than those of any other part of the United Kingdom. Within the London workforce, 60% have Level 4[23] or above qualifications compared with 37% for the England and Wales workforce outside London.[24] London boroughs such as the City, Tower Hamlets, Islington, Westminster and Camden have significantly more highly qualified workers than even university towns like Oxford and Cambridge, let alone the rest of the country.[25]

Migrants not only boost creativity though their own ideas but also through stimulating creativity amongst others. Donald Fan is the Senior Director for the Office for Diversity for the US retailing giant Walmart. He suggests that "diversity of talent, by definition, provides more ideas and perspectives into driving for the best business solutions". He points out that "people see problems and solutions from different perspectives," and that "these perspectives are accompanied by the heuristics that define how individuals search for solutions". His conclusion is that "diverse teams often outperform teams composed of the very best individuals, because this diversity of perspective and problem-solving approach trumps individual ability."[26] Empirical research by Chicago University Professor Ronald Burt[27] backs Donald Fan up: those with more diverse networks solve problems very much more easily than those with less diverse networks.

"As a proportion of its population, and in absolute numbers, London now has a higher inflow of foreigners than do either New York or Los Angeles"[28] pointed out *The Economist* in 2003. 'London Key Facts and Statistics' produced by the Association of London Councils claims that London is one of the world's most diverse cities ethnically,[29] quoting ONS neighbourhood statistics based on the 2011 Census. It also claims that "London has the largest number of community languages spoken in Europe. Over 300 languages are spoken in London schools". Not only did London have a long-established skill base in advertising and marketing, but it had a massive supply of diverse and creative young people, many of them looking for work. Once the impediments of those supereconomies of scale and the network effects described earlier had been overcome, and once online retail and marketing began to take off, London was in prime position towards the end of the 21st century's first decade. It all percolated together and hey presto, we had the Flat White Economy.

CHAPTER 3

What is the Flat White Economy and where is it based?

The UK has developed an enterprise culture with enormous numbers of new businesses since the beginning of the 1980s. When Mrs Thatcher became Prime Minister in 1979, there were only 785,688 companies registered in the whole of the UK. By September 2015 this number had grown to 3,464,155.[1] On another measure, the number of businesses registered either for VAT or PAYE rose from its trough of 2.1 million in March 2011 to 2.45 million in March 2015.[2] London punches well above its weight in its number of businesses with its 13.2% of the UK population, it is responsible for 18.2% of the VAT or PAYE registered businesses and for about a third of the growth in the past 3 years.

It is not surprising that the Flat White Economy has been located in London. The UK government produces statistics on the location of businesses. These show that in 2013 40.3% of all UK information and communications technology turnover originated in London[3] and more than half the growth in this sector in the past 5 years had been in London.

Virtually every tallying of the jobs associated with the Flat White Economy disagrees with the next because of inconsistent definitions. But my colleagues at Cebr have defined the sectors in the FWE as the "MIC": media, information and communications. These in turn comprise:

* Software publishing
* Computer programming and consultancy
* Data processing and websites
* "Other information services"
* Advertising
* Market research
* Television and film post-production

As you can see, the MIC grouping ranges from the very technology-based sectors such as computer programming to more cultural sectors like TV and film post-production. There is a heavy marketing component – a reflection of the fact that the UK was the first major economy to convert from traditional advertising to digital. Internet advertising spend in the UK was £6.3bn in 2013. Total digital spend was £7bn. This comprises 38% of all advertising spend of any kind in the UK. The UK's 38% compares with a global figure of 29.5% according to OFCOM data.

Cebr's research made use of government data and indicated that in 2012 (before the bulk of the growth) the total number of Flat White Economy jobs was 144,500 in London as a whole. One should be very careful when using government data because of the small sample sizes – it means that the numbers for subsectors fluctuate alarmingly. Nevertheless, it became clear that three big sectors made up over

60% of these 144,500 jobs: computer consultancy activities (41,300); advertising (31,000 jobs, up from 19,400 three years earlier); and the catchall "Other information technology and computer services" sector (19,200 jobs). These numbers could well be up by 50% when 2014 figures become available. The Tech Nation Report from Tech City claimed that there were 251,590 'digital jobs' in Inner London alone in 2015 [4].

The patch of London we are talking about is known by a number of names – 'Silicon Roundabout', 'Tech City', 'East London Tech City'. The boundaries are ill defined and have varied over time but the conventional definition currently is that the area extends from Old Street (the boundary between Central and East London) in the West to the Olympic Park in Stratford East. It is claimed in Wikipedia to be the third-largest technology startup cluster in the world after San Francisco and New York City.[5] Some of the digital technology world's major companies – most famously Google, Facebook and Wikipedia – have invested in the locale. A range of academic institutions have too, including Imperial College, the UK's leading scientific academic institution, who have invested in a major £1 billion campus there. According to *Wired* magazine, the number of technology companies in the area rose from 15 in 2008 to 85 in 2010, and from 200 in 2011 to over 5,000 in 2012. But this only takes into account businesses that are fully formed and in operation. According to Tech City in 2015, 40,000 digital businesses are based in Central London. The highest density (according to estate agents Stirling Ackroyd) is in EC1V with 3,228 tech firms per square kilometer. The second highest density is EC2A with 1,520 tech firms per square kilometre.[6] Both local and central government realised that they had a success on their hands and have rushed to associate themselves with it.

Prime Minister David Cameron involved himself at an early stage in his premiership. The boom in the area overlapped fortuitously with the build-up to London's hosting of the 2012 Olympic Games. In November 2010 Cameron stated: "Our ambition is to bring together the creativity and energy of Shoreditch and the incredible possibilities of the Olympic Park to help make East London one of the world's great technology centres".[7] It was an initiative from Cameron to christen the area 'Tech City' and at the same time set up a promotional agency of the same name. Cameron's Senior Policy Advisor at the time, Rohan Silva, took the development of the area very seriously; so much so that he quit his job in Downing Street to become involved in various local enterprises, including Spacious, a website (and now app) that facilitates the finding of office space. Four years later, David Cameron opened the Technology Exhibition CeBIT in Hanover with a follow-up of sorts: "Come over to Shoreditch in East London and you can see it – Tech City is teeming with startups and new ideas. It started less than three-and-a-half years ago with 200 digital companies in that area of East London – now there are 1300".[8]

London Mayor Boris Johnson has been even more directly involved. He has announced as his aim the desire to make London the technology capital of the world. Speaking at the launch of London Technology Week in June 2014, he pointed out Tech City's successes but bemoaned the lack of blockbuster take-offs: "Although we've got the biggest tech sector in Europe we haven't yet produced knock-out multi-billion pound companies. They have in Silicon Valley. We need to explore why that is. Is it a certain British diffidence about making billions? Is it that we haven't got kick-ass business people here? Or that the banks aren't as proactive as they should be?" Boris Johnson has a point – the Flat White Economy has produced a lot of startups but they have not yet grown to the scale of global leaders yet. There is no London-based Facebook, Google or Alibaba. This may be partly a matter of time, but it remains a valid criticism.

There are however three Flat White companies that have now reached 'unicorn' status – the local jargon for having a market capitalisation of a billion dollars or more. These are Shazam, a music identification service, TransferWise, described as the Skype of money transfers and Farfetch, an online fashion marketplace.

Besides the digital businesses in the area that directly comprise the Flat White Economy, there are related businesses which, in the decentralised world that many technology businesses inhabit, often involve more jobs and activity than the core businesses themselves. There are also 'downstream' economic effects from the induced consumer spending. Some analysts include them in their estimates of economic impacts while others do not. But whether they are included in the analysis, they definitely take place in the real world! The "upstream" spending on associated business services are very much part of the economic impact of the digital economy. One of the major impacts is through the requirements for venture capital for the many business startups in the area.

Financing technology

With the City of London on the doorstep of the Flat White Economy, the potential for funding is positive. And with the traditional banking system still inhibited by the aftermath of the financial crisis, which has held back its ability and willingness to lend, there is a niche that needs filling.. London accounts for 8.9% of world technology finance – a long way short of the 30% contributed from Silicon Valley but nevertheless a figure which indicates an emerging growth. In the five years to 2013 the growth in investment in financing technology in London was twice that of Silicon Valley. The UK and Ireland (which is, in practice, chiefly a London measure) accounts for more than 50% of European technology venture capital measured by number of deals – and 69% measured by funds invested. Deal volume in London has been growing at an annual rate of over 70% since 2008 and doubled in 2014. The conjunction of the Flat White Economy and London's financial hub has meant that a key driver of digital jobs in London is Fintech.[9] It is claimed that London is the Fintech capital of the world with 44,000 employed in the sector – very slightly more than in New York. London is now home to at least five fintech laboratories: the Fintech Innovation Lab set up by Accenture in 2012; the Level 39 Fintech accelerator in Canary Wharf; the Techstars/Barclays lab in Mile End; the StartUpBootCamp backed by Lloyds, Rabobank and MasterCard; and Bold Rocket in Great Eastern Street. These all provide a mix of finance, training and support for high-tech startups.

So we can see why London is becoming such an exciting hive of digital activity. But why East London in particular?

Maps prepared by the Corporation of London[10] showing how the geographical sector was defined officially in 2012 and reveal an area essentially centred on the City Fringes. As early as 2007, Cebr identified for the Corporation of London a potential symbiotic relationship between the City and its fringes. The City Fringes are defined as the boroughs neighbouring the City of London – Camden, Hackney, Islington, Lambeth, Southwark, Newham and Tower Hamlets. In 2007 they all had unemployment substantially above the London average.[11] Moreover, all these boroughs were in the top 25 most deprived of the 354 boroughs (or equivalent) in the UK; this was measured by using the government's Index of Multiple Deprivation.[12] But even then these areas benefited from £1.8 billion of spending from businesses and organisations based in the City of London, with advertising being the largest single area and market research, computer services and auxiliary financial services also contributing.

This proved to be the basis of the burgeoning digital economy. One of the driving forces behind the location of the Flat White Economy is where young people live. Other than Bloomsbury, where the proportions are inflated by London University, the areas with the highest proportions of young people (18–29 year olds) as residents in London are: Finsbury Park, virtually all of Tower Hamlets (especially Millwall, Blackwall, Bow East, Cubitt Town, Whitechapel and Bethnal Green) and parts of Camden, Hackney and Haringey. South London from Wandsworth through Lambeth to Southwark also have typically nearly a third of their residents in the 18–29 year old category.

Although there are many digital companies located in the high-rent areas of the City of London and Westminster, these will tend to be companies that service high-value clients. Rents in these areas vary, depending on the economic cycle, between £50 to £80 per square foot. Typically a single employee needs 125 square feet of office space, which includes space for services. So City of London office space costs around £8,000 per annum per employee. This makes economic sense when the employee is getting paid £200,000 or more – any marginal improvement in productivity brought about by convenience of location is worth paying for. In any case, the City business model is centred upon the success of a few employees with high productivity, so the space requirements are low as a proportion of the total value added – and therefore high rents are affordable. But the business model for the Flat White Economy operates via larger numbers of relatively low-paid employees – they might typically earn £25,000 to £35,000. This means that space is a significant cost-driver and so employers look for relatively cheap accommodation.

When I moved my office into EC1V in 2000 the typical cost of property was £15 a square foot for rent – less than £2,000 per employee. This meant that startups could afford to rent sufficient property to allow room for expansion without the cost becoming exorbitant. The Old Street area became the epicentre of the Flat White Economy for four reasons. Property was (initially at least) relatively cheap. There were good transport links: it was accessible by both public transport and bicycle routes to the areas in North,

East and South London where many young people live. There was already a core of relevant telecoms, IT, marketing, advertising, financial and film businesses based in the area. And, lastly, the area had a good reputation for the arts, clubs and for trendy bars. At one point Shoreditch was said to have the highest concentration of art galleries in the world as well as a plethora of clubs and trendy bars. There are 38 clubs within 300 yards of Old Street station (including London's first dedicated e-cigarette bar).

None of these factors on their own would have been sufficient to make the Flat White Economy end up where it did. But together they became synergistic and determined the location. Then the area became fashionable. Not only did the shopping and leisure facilities improve but the area became a focal point. People would come into the area for coffee during the day or to drink and dance in the evenings. As it acquired a fashionable veneer, employers found that they could attract more and better employees by being based around Old Street and into Shoreditch. Reputation, of course, builds on itself. By the time the government had given the area its seal of approval, calling it Tech City and setting up the Tech City organisation on the Mile End road, it was already a success.

CHAPTER 4

Flat White Lifestyles

What makes the Flat White Economy tick is arguably the number of people available to work in it. Many of them are attracted to London by jobs – London has been the only place in Europe creating jobs for young people at Western rates of pay on a significant scale over the past ten years. And the virtual collapse of employment opportunities for young people in most of the Eurozone has stimulated mass migration into London.

But the people aren't only attracted by the jobs. Often they come to London to look for fun. Once they are in London, of course, they then look for work to pay for the fun. This creates the supply of skills that support either the Flat White Economy itself with high-tech skills know-how or the support industries that rely on more prosaic abilities like making coffee or mending bikes.

A characteristic of Flat White Lifestyles is mixing. Work is mixed with play. Living is mixed with working. The coffee bar is mixed with the home and the office. Some of these lifestyles were pioneered in Silicon Valley but have mutated in the close-knit high property cost environment of East London.

Back in the 1980s and 1990s, Londoners distinguished themselves with their appetite for champagne. During the heyday of financial services, UK imports of champagne (not all sold in London but mostly so) rose four times. The rise of the flat white economy has been associated with a different lifestyle. The bicycle has replaced the Porsche, skinny jeans have replaced suits and, of course, flat white coffee has replaced champagne. Champagne sales are down a quarter since their peak in 2007. Now there are 3.2 million cups of coffee sold in London every day, an increase of more than 50% since 2007.

Twenty years ago the tone for London living was set by the 'Loadsamoney' style of the rich bankers of the 1980s and 1990s. But it was extravagant, elitist and generally failed the test of good taste. Those working in financial services earned much more than they knew how to spend and had a very boyish nouveau riche view of how to spend their money. Although they set a tone, driven by their massive spending power, the lifestyle itself was expensive and not easily copied.

The Flat Whiters are the new style leaders. Their salaries are low so they have to make up with their brains for what they lack in buying power. East London is now one of the world's desirable shopping locations with quirky boutiques selling unusual clothing and other items. The Boxpark on Bethnal Green Road in Shoreditch, for example, is according to the *Daily Mail* the world's first 'pop-up mall', featuring fashionable stores and bars with a design picking up on the unique atmosphere of urban lifestyle, fashion, art and design in this now trendy area. And it is much easier to copy the trends set by the Flat Whiters than the 'Loadsamoney' types from financial services in the 1980s and 1990s – their spending patterns are affordable. The only difficulty is in keeping up with the trends. Flat Whiters know that they can't price their styles out of the market so to keep ahead their styles have to keep changing.

Coffee Shops

And of course the most characteristic service businesses in the Flat White Economy are the coffee shops. When the *Guardian* newspaper set out to find London's top coffee bars, six out of the top ten were in East London[1].

Coffee shop sales are increasing 10% annually in London at present[2] – no mean feat when average earnings are rising by less than inflation. Coffee seems to be the new beer and coffee shops the new pubs, filling the same social role. Even outside London, coffee sales are growing by about 7%. Nationally sales now amount to £6 billion. The market which includes branded coffee chains, independent coffee shops and non-specialist operators, grew to 15,723 outlets in 2012, up by 4% over the past year.[3] The British were once a nation of tea drinkers, but more and more people are becoming coffee drinkers. This mirrors the transition occurring in the London economy and society.

Allegra Strategies are a strategic and retail consultantsancy who have carried out a number of studies of the retail coffee shop sector. They predict that the total UK coffee shop market will exceed 20,000 outlets and £8 billion turnover by 2017, driven by branded coffee chain expansion and non-specialist operator growth.[4] The biggest chain, Costa Coffee now owned by Whitbread, has

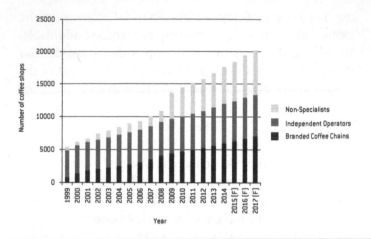

Figure 4.1: Total number of UK coffee shops by outlet and by type with forecasts[6] (Source: Allegra Strategies Ltd)

1,831 branches[5] with nearly 80 new openings in the first half of 2014. It expects to have 2,200 branches in 2018.

Coffee shops represent both a statement of lifestyle for the Flat Whiters and a focal point for creativity. Trendies pride themselves on being coffee connoisseurs, while the coffee shops with their free Wi-Fi and open table arrangements act as ideal meeting places for the creative processes required by Flat White industries. Throughout the day and much of the night, the sight of young people absorbed in silver MacBook screens besides a paper cup is common throughout the East End.

Flat Whiters often have fairly cramped accommodation. So the coffee shop is their study, their drawing room, and their dining room. It can even be their bike repair shop – cycle cafes such as 'Look Mum No Hands!' (whose original branch is located on Old Street, one of three such cafes which opened in a single week in May 2010)[7] have become a widespread phenomenon in London, often busy from breakfast until closing at 10pm and offering free Wi-Fi and (not free) coffee in addition to services such as bike-washing, parking and repairs.

Andy Harrison, chief executive of Whitbread, believes coffee shops fill a hole in British society once met by pubs. Families and women in particular use coffee shops for social gatherings, he says. He also points out that women now have greater spending power than in the past. "Think of the coffee shop as a social venue," he says. "What we have seen is the [UK] coffee shop market has grown at about five per cent per annum throughout the recession even in the most economically challenged parts of the UK. We think the reasons behind that are to do with things like the growth of female independence, female spending power. Over half of our customers are women. People talk about the pub as a meeting point but pubs were more about males and the evening, coffee shops are [open] all day, more female [orientated] and certainly more family." [8]

He also believes coffee shops have been boosted by people shopping more online. Instead of spending their Saturdays trawling the shops, Britons are meeting up with friends at their local cafe. By comparison, the number of pubs in Britain has been in steep decline for a number of years. Sales of beer in pubs declined by 41% between 2000 and 2013. The decline in the number of pubs has been equally dramatic, with a fall from 60,400 in 2000 to only 48,006 in 2013.[9]

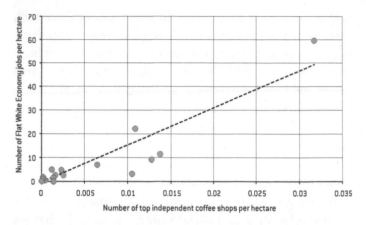

Figure 4.2: Relationship between Flat White Economy jobs per hectare and number of independent coffee shops per hectare. Correlation in fact is slightly weaker when more specific employment data is used – but the correlation is still strong.

My colleagues at Cebr have even done a simple economic analysis looking at the number of independent coffee shops per hectare for London boroughs compared with Flat White economy jobs per hectare. The coefficient of determination R^2 (which measures the relationship between the two sets of data) turns out to be 0.845, which suggests a very high correlation. Which causes which, is not quite so clear: the relationship is shown in Figure 4.2.[10]

Bicycles

The most characteristic vehicle for Flat Whiters is the bicycle. The first ever proper survey of cyclists in 2013 found that 49% of all vehicles travelling along Old Street (the epicentre of the Flat White Economy) during the morning rush hour were bicycles.[11] There is no comparative historic data but – having worked in the area for fourteen years – I would be surprised if ten years ago the proportion was as high as 10%.

The 2011 UK census failed to find any evidence of a significant increase in cycling for commuting purposes in London but for those who use the roads regularly, such a result is clearly nonsense. There is hard data from the TfL cycle survey which shows that for the twelve months to April 2014, on the TfL monitored routes (which are meant to be a representative sample but may not be) the number of cyclists was 3.4 times the number in the base period of measurement in 2000/01.[12]

Sales of bicycles in the UK have been running at 3.6 million a year for the three years to 2012 – strongly up from the 2.3–2.4 million which they averaged during the 2000 to 2002 period.[13] The number of bike shops in the UK has risen by 15% in the past ten years.[14]

The weather in London is seen as a discouragement to cycle usage, but actually it's not significantly worse than in Amsterdam,[15] which is generally seen as the European cycling capital. In Amsterdam 60% of journeys in the inner city are made by bike and 38% of journeys overall.[16] More of a discouragement for cyclists is the danger of cycling with roads full of cars and lorries travelling at variable speeds (though some Flat White adrenalin junkies even seem to be encouraged by the challenge to life and limb).

Much attention was paid when six London cyclists were killed in two weeks in November 2013. For the UK as a whole 109 cyclists were killed in 2013 which may suggest that the risk is low, but 3,143 were seriously injured which suggests a rather greater risk. I myself have my front teeth embedded in Upper Street, Islington as a result of a cycling accident – maybe if I ever get famous there will be a blue plaque on the road – but fortunately I survived with a slightly uglier face!

Cycling in London is statistically becoming safer. According to the *New Scientist* "The annual KSI, or average number of cyclists killed or seriously injured, between 1994 and 1998 was 567, compared with 515 per year from 2008 to 2012. That's a fall of nine per cent. At the same time, there has been a huge rise in the number of people getting on their bikes – 176 per cent more since 2000."[17] But the significant danger of at least serious injury must be a discouragement to many potential cyclists.

Clothing

According to Business Insider, Peter Thiel, founder of PayPal and one of Silicon Valley's key venture capitalists, "hates suits". While he cautions that there are "no absolute and timeless sartorial rules," Thiel says that, "in Silicon Valley, wearing a suit *in a pitch meeting* makes you look like someone who is bad at sales and worse at tech." Maybe that's why he has a simple rule for investing: never bet on a CEO in a suit. Thiel says that this rule has helped him avoid making poor bets on slick business folk compensating for crap products with well-dressed charm. "Maybe we still would have avoided these bad investments if we had taken the time to evaluate each company's technology in detail," Thiel writes in his new book. "But the team insight – never invest in a tech CEO that wears a suit – got us to the truth a lot faster."[18] The Flat White dress code is similar to Silicon Valley. Jeans, trainers, hoodies, bomber jackets – anything but suits!

In recent years the East End of London has transformed into a fashion hub. Where historically it was the site of the textile factories that supplied the materials to smart stores in Knightsbridge and Mayfair, many of the old factories have been reinvented as post-season discount outlets, such as the Burberry factory outlet at Chatham Place.

The industrial feel associated with the East End promotes an edgy vibe which attracts Flat Whiters, further cementing the area as synonymous with indie creativity and innovation. Notable designers such as Wayne Cooper and Alexander McQueen grew up in the East End and three locations in the EC postcode area were selected as venues for the 2014 Spring/Summer London Fashion Week, adding to the area's reputation for bohemian style.

Work and play

The UK *Esquire* article about Silicon Roundabout[19] describes a 'Ping Pong Fight Club', "a wholesome night-time gathering of the brightest tech sparks from London's Silicon Roundabout, and a mock-competitive midweek tournament featuring 64 players (both male and female, though mostly male), sixteen rounds of table-tennis, six bright blue tables, four sassy semi-clad cheerleaders (courtesy of Crystal Palace FC), a vast array of wacky shorts, T-shirts and even wackier headbands, and a singularly woeful MC whose repeated comedic references to early Nineties Saturday TV staple *Gladiators* fall on the deaf ears of an audience whose mean age is pushing the upper limits of 27."

The point is made that the Silicon Roundabout lifestyle involves both work and play and that the two are mutually reinforcing. The article describes the teams as representing a range of different types of local businesses: "The ping pong teams thus represent the area's so-called scrappy start-ups, including Decoded, (learn coding in a day!), Songkick (get concert alerts on your phone!) and Funding Circle (get money for small businesses!), as well as heavy-hitters such as Mind Candy (makers of the popular online kids' game Moshi Monsters) and behemoths such as Facebook. Although it's not as if the event was ever going to be short of players."[20]

An evening like this is half-work half-play – many conversations revolve around recruitment. Such "event management systems" serve the dual purpose of socialising and networking, with the line between the two increasingly blurred. It seems multitasking is a hallmark of the new hyper-productive industry that mimics the type of lifestyle developed by Silicon Valley employees, where the value of play is considered catalytic to creativity. According to *Esquire*, some companies provide up to 20% time off for employees to pursue personal projects.

Meanwhile work and play is also mixed in offices which have playrooms and cushions for chilling out. The headquarters of Mind Candy and Google feature conspicuous multi-floor slides, justified on the benefit they provide in channelling a creative and innovative environment. In an industry centred on youth, the value of 'chilling' it seems cannot be underestimated.

Google have started work on a new London headquarters office set for completion in 2016 in Kings Cross. According to the description in *The Guardian* : "Google's spanking new £1 billion London Headquarters [is] designed to make going to work feel like an exciting day out."[21] The office is equipped with a climbing wall, indoor football pitch and a rooftop swimming pool. The architect is Simon Allford of Allford Hall Monaghan Morris who commented: "The idea is that the people who are in the building – not the tenant but the actual staff – need to be attracted to the building. They need to like the community of the building."[22]

The new building will stretch over 300 metres along a new street carved out between Kings Cross and St Pancras stations and will apparently look like, "a vast ocean liner run aground".[23] The building is scheduled for completion in 2016 and will house approximately 750,000 square feet for 4,500 staff. There will be a snaking ramp allowing employees to cycle all the way into the building. Allford describes the building as a theatre suitable for the company's self-image, "as a dynamic, open network in constant flux". The building the article reports will therefore will be a flexible framework, "to be inhabited by a changing landscape of beanbags and thinkpods, laptop stations and the occasional static desk".

The reason for structuring the office environment as a hybrid of youth club and holiday camp, says Google, is to "encourage casual collisions of the workforce". As they bump into each other on their micro-scooters and splash around in the pool, employees will always have a laptop or tablet on hand nearby, for the creative encounter to be channelled into monetisable productivity. Something of this outlook is satirised in the novel *The Circle* (2013) by Dave Eggers where all the pressure for creative socialising is taken to even greater extremes at a corporate campus in California with more than an echo of Google's. This hybrid mix of home and office is very much the style for the Flat White Economy.

Electronic devices

One of the defining characteristics of the Flat White Economy is its embrace of information technology. The productivity of the Flat Whiters is supercharged by the latest gadgets, which offer everything from classy design software to the unprecedented economies of scale available through the internet.

Flat Whiters typically own at least one of the latest smartphones, a tablet and a high-tech notebook. If they are doing well, they probably own top of the range sound and other equipment as well. Most likely, these products form a large proportion of a Flat Whiter's otherwise frugal expenditure (the 64GB iPhone 6 when introduced was a week's median income for people living in Hackney and Islington),[24] which serve the dual purpose of workplace necessity and style statement. Among this group, the popularity of design-centric products such as those from Apple and Beats not only allow them to tap into the fast digitizing industry but also serve as signal of style and identity. The Porsche or Ferrari and the Rolex watch have been replaced by the MacBook Pro and the latest iPhone.

In 2012 mobile and social media app agency We Are

Apps found that 70.5% of men and 77.5% of women, aged 15–34 respectively, in the city owned a smartphone of some form. The data shows that ownership of smartphones may not even vary proportionally with income – for example, marginally more lower-middle class male Londoners aged 24–35 owned a smartphone compared to upper class male Londoners of the same age, 81% compared to 80% respectively.

Property and location

The feature that most people associate with London property is expense. Flats and houses in London can be expensive as anywhere in the world and since salaries in the Flat White Economy are not generally high, in some ways it is surprising that this economy has located itself in London.

Flat Whiters offset this in two ways. Firstly, they tend to live in the relatively cheaper parts of East London (although as the Flat Whiters gentrify these areas, prices rise). And secondly, they economise on living space.

The travel to work data from the 2011 Census shown in Figure 4.3 (overleaf) shows where the people who work in Hackney and Islington live. Of the 223,438 people recorded in the Census as working in the area, 50,000 or between a fifth and a quarter, lived locally in the two boroughs. Roughly half of those who lived locally travelled to work on a bike or on foot. The rest commuted from a range of boroughs or from outside London, though nearly 14,000 commuted from Haringey.

Because property values even in East London are kept

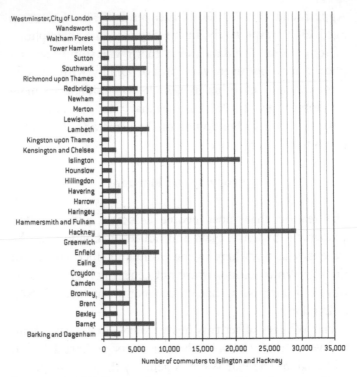

Figure 4.3: 2011 Census data on origins of people commuting to work in Hackney and Islington. Source: www.ons.gov.uk

high by supply and demand, many Flat Whiters share accommodation. The lack of domestic living space is one reason why they make such use of cafes and why employers are encouraged to provide play areas. In addition, lack of space affects ownership of possessions – there is little space to store possessions so Flat Whiters tend to have a backpacker's range of possessions – relatively few!

Pubs and clubs

Central London remains the epicentre of London's clubs and pubs. According to the recent study in February 2014 for the British Bars and Pubs Association,[25] the pubs and bars in the Cities of London and Westminster contribute £584.4 million annually to the UK's Gross Value Added (GVA) and over 15,000 jobs directly and indirectly.

But the growth of the Flat White Economy has been associated with a new burgeoning nightlife scene which is now London's second most important area. Taking Islington, Hackney and Shoreditch together they jointly contribute 5,645 jobs – a third as many as the Cities of London and Westminster – and £252.6 million to the UK's GVA – in other words 35% as many jobs as the twin Cities and 43% as great a contribution to GVA. Given Central London's historic and transport advantages, these numbers are highly significant as an indicator of the takeoff of East London as a centre for partying.

Meanwhile the decline of the impact of the City and financial services can be seen in the tracking of UK sales of champagne, which doubled in the 1980s to 20 million bottles a year.[26] They fell back to fourteen million in 1991 in the 1990s financial crash but recovered to a peak of 39 million bottles in 2007. Since then they have fallen back in five of the past six years and in 2013 were down by nearly a quarter from the 2007 peak at 30.7 million bottles[27] – although the UK remains the world's largest export market for champagne.

So a new lifestyle is emerging in the East End of

London – more modest than the style of the rich bankers but much more self-consciously stylish with rapidly changing fashions and with work and play mixing. It is nothing like the ferocious and expensive lifestyle of the financial service workers of the 1980s and 1990s but is much more modest. It is much affordable to copy, though the pace at which styles change mean that one has to put a lot of effort into staying hip.

CHAPTER 5

Copying the Flat White Economy

Obviously in London the Flat White Economy is a dynamic force which has already made a major contribution to the UK economy (for more see Chapter 8). But how can its contribution be enhanced and how can it be turned into a more widespread phenomenon that benefits the whole of the country? Many local authorities around the world, as well as in the UK, have tried to do this and have failed – there seems to be something about trying to engineer a digital technology hub that often eludes policy makers. Yet there must be lessons that can be learned from looking at what drives technology clusters around the world – this chapter tries to identify them. I have looked at the US in some detail and also at major technology clusters around the world to try to see what appears to work and what doesn't. Do they offer lessons that make it clearer as to how alternative centres to London might be developed in the UK?

The US

Every year in the US there is an awards ceremony for the top Digital Cities run by the Center for Digital Government and Digital Communities programme. The Center is a national institute focusing on research and advice on information technology policies and practices in state and local government. It is a division of e.Republic, which is a media and research company specialising in state and local government and education. The awards are sponsored by heavyweight IT companies such as AT&T, Sprint, Laserfiche and McAfee.

Although the awards are based on what the various levels of government have achieved rather than on the emergence of private sector activity, the awards for 2013 do seem to have hit targets where both are in evidence. The awards highlighted about fifty different localities that could present themselves in some way as tech hubs across the US. Top in the over 250,000 population category was the City of Boston. Irving, Texas won the 125,000–250,000 population category and who else but Palo Alto (of course) won the under 75,000 category?[1] In 2014 Los Angeles headed the 250,000+ category, Winston-Salem in North Carolina the 75,000–125,000 category and Dublin, Ohio edging out Palo Alto in the under 75,000 benchmark.

Using a different methodology based on the growth in science, technology, engineering and mathematics (STEM) based jobs, *Forbes* magazine in 2012 undertook a different assessment of the US's key digital communities.[2] The list identified by Forbes lists the following as hubs:

- The Seattle metropolitan area that has seen 12% tech job growth and 7.6% STEM growth from 2001–11.
- The Washington-Arlington-Alexandria area to the south of Washington DC which has experienced more than 20% STEM job growth and has amongst the highest base proportion of tech jobs overall.
- The San Diego area, where tech jobs have growth by 30% over a ten-year period.
- Salt Lake City, also with 30% growth where the relatively inexpensive property is a major attraction.
- The Baltimore area to the North and East of Washington DC with 40% growth in tech jobs up until 2011, though the large squeeze on Federal spending will have reversed this trend.
- Jacksonville, Florida, though this is within an area noted most of all for growth (70% in a decade) from a low base.
- San Jose-Sunnyvale-Santa Clara – clearly Silicon Valley couldn't be left out, though its tech job growth has slowed down from a high base.
- Columbus, Ohio has both a high share and a rapidly rising number of tech jobs which appears to have accelerated in recent years and which is based on private spending rather than the Federal government.
- Raleigh, North Carolina has traditionally been a research centre with its Research Triangle. (When I was an IBM employee, as long ago as 1986, I was taken round the area by a local North Carolinian friend with whom I had been at university and the area was already doing well). It has also had now more than 30% tech job growth in the past ten years.

- Finally, the area around Nashville, Tennessee is no longer just a music-making hub. It is spreading its wings into the tech sector with 43% tech job growth in the past decade.

The analysis in the Digital Cities award listings and in the *Forbes* listing suggests various tentative conclusions. Firstly, a digital track record can be a stimulus to future growth but isn't absolutely essential. Secondly, a base like Nashville that already has some relevant Flat White Economy ingredient like music has a good chance of becoming a digital hub. East London already had a significant cultural presence in film and arts before the Flat White Economy emerged.

One factor that often helps the sector develop is the availability of cheap, relatively highly-skilled labour. Often areas which have relatively cheap accommodation attract such people and help provide a base. Clearly inexpensive accommodation is not a characteristic usually associated with London! But the need for relatively cheap accommodation is probably the main reason why the digital industry in London has developed in East London rather than in West London.

Worldwide

MIT Technology Review[3] has identified eight technology clusters around the world with sufficient critical mass to make progress – Silicon Valley of course and Boston with its leadership in electronic medical applications, Tech City/Silicon Roundabout in London, the Paris-Saclay

Cluster, Moscow Skolkovo Innovation City, Israel, Bangalore and Beijing.

MIT identify strong intellectual property protection, liberal immigration rules (see Chapter 7), an entrepreneurial culture and government support as key elements making for the success of such clusters. They also suggest that a warm and pleasant climate helps. From my own experience in the sector I would add market access as another necessary condition. Although established businesses can sell at a distance, the majority of new businesses need face to face contact at the current stage of technological evolution, though this could well change in the future (perhaps in ten years' time or so as video-based conferencing matures).

Although cheap accommodation and office premises are important drivers of the initial stages of emergence of a tech cluster, once the cluster has acquired critical mass, rents tend to rise which in turn constrains growth. In Kendall Square just off the MIT campus in Cambridge, Massachusetts rents have reached $70 per square foot. According to *Technology Review*, "Amazon has moved a mobile development team to the area, Google has expanded quickly into new buildings and drug companies are piling in too."[4] Around Silicon Roundabout in London, rents shot up in 2013 to over £40 per square foot (£70 including services), but have fallen back slightly since as new space has been attracted on to the market.

In Paris, Moscow and Beijing the drive to replicate Silicon Valley is strongly supported with government money. Yet economists argue about whether it is possible to promote technology top down. On the one hand, the original development of Silicon Valley had much to do with US government defence spending and the current development of the Israeli technology sector is again driven by military spending. In addition, in China, technology giant Huawei allegedly grew out of the communications arm of the People's Liberation Army, while the Russian technology cluster also appears to have had historic links with military developments. On the other hand, the development of the Flat White Economy in the area around Old Street has had relatively little to do with direct action by either central or local government[5] and much more to do with other factors such as lifestyle, property prices and the chance factor of the collapsing Eurozone economies in Southern Europe sending their best and brightest young people to work in London.

Paris–Saclay

When he was French President in 2006, Nicolas Sarkozy launched the Paris-Saclay project with the aim of creating a cluster of world-class universities, research facilities, science parks and technology firms that could compete with tech hubs such as Silicon Valley. It is overtly modelled on the image of Cambridge, Massachusetts, and the €5 billion euros pledged by the government on the project is planned to fund everything from individual projects to an infrastructural network: €2.5 billion will go into building projects for teaching, research and innovation; €1.3 billion will be spent on learning, technology and science programmes; and €1.2 billion will go into establishing a transport network that plans to add another line to the Paris Metro.[6] In addition, an estimated €15 billion of private finance will go into establishing facilities for student accommodation, business and research. One of the stated objectives of the Paris-Saclay project is to attract to the area some 25,000 more employees, including 12,000 researchers.[7] The region already contributes 15% of French industrial research and development (R&D).[8]

The Paris-Saclay project, almost the creation of a mini-city, has many of the ingredients for success in developing a digital cluster. The billions of private and public funds invested into the area is likely to result in the establishment of world-class research facilities in the vein of France's impressive existing science and technological base, and a good transport network plus proximity to Paris means that such facilities can easily tap into services provided in the capital. In addition, an impressive concentration of research is already based in the Saclay plateau, one of the highest such concentrations in Europe.[9] Also, with the relocation of many of France's grandes écoles (elite French higher educational institutions that students have to undergo special curricula for years to enter) into the area, there will be no shortage of exceptional talent. The University of Paris-Saclay took its first students in September 2015 and is aimed at having the critical mass of Oxford, Cambridge or Harvard.

But will it become a major contributor to the French economy in the same way that the Flat White Economy is contributing to London? There are clear similarities with Cambridge, Massachusetts, where tech giants such as Google, Amazon and IBM (who has an innovation centre there) are located. These firms use the area to concentrate talent into innovation and R&D in special facilities. The more industrial elements of R&D are suited to such facilities, though typically in the more mature sectors of particular industries.

The more creative end of the industry is perhaps less likely to relocate to the region – although the extensions of public transport will improve accessibility, as the location is difficult to access without a car. People are, perhaps, unlikely to want to live nearby for a long time to come. And although about 20% of the space available for the private sector has been occupied or will soon be, the main occupant is going to be the French electricity giant EDF who will move into the area during 2016.

Paris-Saclay will probably succeed eventually but it will take time. It will be held back by the weakness of the French economy and the slow progress towards digitalisation. And the area is more likely to be suited to the industrial end of R&D than the Flat White Economy sort of R&D. But as students move into the area its character is likely to change and it is hard to see this area failing to succeed, even if progress is slower than might be hoped for.

Skolkovo Innovation Centre

Skolkovo was a pet project of former Russian president and current Prime Minister Dmitry Medvedev. The Skolkovo Innovation Centre is about two kilometres southwest of the Moscow ring road and is an attempt to mobilise Russian science and technology to boost its tech industry performance. It is similar to the other tech hubs in the world, with a focus on 'clusters' of information, energy, nuclear, biomedical, space and telecommunications technology. It is also designed as a mini-city, with a hospital, school, apartments and offices all available on-site. It is extremely ambitious, with 500 billion roubles of the federal budget planned to be spent on the project from 2013 to 2020[10] with the aim of the project being fully operational by 2030.

The project is an attempt to stimulate other areas of the Russian economy and reduce the dependence on natural resource extraction and oil revenue. It has some similarities to the way in which the Flat White Economy supported the London economy when the financial services sector weakened in 2007. By providing attractive tax breaks the centre aims to house a high concentration of startups, and it has already attracted big-name partners such as IBM, Intel, Cisco, Siemens, Boeing, Nokia and Microsoft. An article in the *Moscow Times* has claimed that it could contribute 1.5 trillion roubles to the Russian economy by 2030.[11]

Since its conception, however, the project has been plagued by scandals and allegations of corruption, though there have been no major prosecutions yet. And as Prime Minister Dmitry Medvedev's star has waned, development has slowed. President Putin has cancelled an order of $910 million that was to have been contributed to Skoltech by state-run companies and overruled a law that exempted Skolkovo developers from obtaining state planning permits.[12] The project also lacks support from universities or the regional government, and has been criticised for being primarily a public relations (PR) project with lofty and ill-executed parameters.

Such controversies could hinder Skolkovo from gaining momentum and getting off the ground, threatening to turn it into another white elephant. Moreover, the weakening growth prospects for the Russian economy are putting the budget for the project under severe strain. Yet the innovation hub is growing, with more than 1,000 companies now based there, and successes such as Aerob, producing surveillance drones for the Russian Defence Ministry,[13] are likely to stimulate other investment. In addition, cooling relations between the Russians and the rest of the world are likely to promote increased self-reliance which may paradoxically further help the project develop.

One suspects that the project may become yet another arm of the Russian military-industrial complex rather than the international digital hub that was originally intended.

Israel

As an innovation hub, Israel possesses some impressive statistics. It has the highest number of patents per capita in the world, its R&D expenditure as a percentage of GDP is the highest in the world,[14] and it is a world leader in information and telecommunications technology.[15] The year of 2013 was certainly a vintage one – acquisitions of Israeli high-tech firms totalled $7.6 billion, including the purchase of Waze (a popular road navigation app used by almost 50 million drivers worldwide)[16] by Google for nearly a $1 billion.

Israel has historically been a centre for technological innovation. The first flash drive was invented in Israel as well as the world's first internet messaging service. Since opening an R&D centre in Haifa in 1974, the Intel development centre has developed crucial computer components such as the Centrino chip and dual-core processors.[17] Since then firms such as IBM, Cisco, Google, Microsoft, Qualcomm and even Apple have developed significant presences in the country. I have been the keynote speaker for the past five years at the Power 50, the world e-gaming industry conference, and have discovered that most of the software driving the e-gaming industry was developed in Israel, generally as a spin-off from the Defence Ministry.

Many factors contribute to the success of Israel as an innovation hub. It is a small, but highly networked and highly educated country with a strong sense of creativity and entrepreneurship. Its unique geopolitical situation also seems to foster a necessary culture of innovation in the face of daunting opposition – much of the reasons for Israel's strength in telecommunications technology, security and encryption stems from substantial funding by the government for military purposes. It also has an exceptional multicultural and multilingual talent base and Jewish mass migration from Eastern Europe in the past quarter century has massively expanded its skills base.[18]

In 2015 the sector even managed the significant victory within the Israeli bureaucracy of achieving a relaxation of Israel's notoriously rigid immigration rules for 'innovation visas'. Despite the geopolitical problems of the Middle East, the Israeli tech sector has such momentum that it is highly unlikely to stumble unless Israel gets involved in a war on so large a scale that it makes it impossible for the industry to operate.

Bangalore/Bengaluru[19]

Known as 'India's Silicon Valley', Bangalore is the focus of the IT and software industry in India. IT services contributed 7.5% of India's GDP in 2012. In the MIT review Bangalore was listed as one of the top eight tech clusters in the world, though MIT seemed to find it hard to explain why it had developed so successfully. Indeed, the main conclusion of the MIT analysis seemed to be that Bangalore's success was due to good weather! In fact, Bangalore is roughly 3,000 feet above sea level and is the highest major city in India, so its climate is generally seen as cooler and much more pleasant than that of many other parts of India. Actually, it is not otherwise difficult to find good reasons for Bangalore's success.

The key point is that this did not happen overnight. It started as early as 1976 when the Karnataka State government (Karnataka is the state of which Bangalore is the capital) planned the first technology park in India. The industrial park, Electronics City, was founded in 1977 on 335 acres of land. During the 1980s software and systems integration businesses led by Wipro and Infosys, set up in Electronics City. Growth in the 1990s followed the rise of the internet.

Today, Bangalore is home to well-known Indian based companies including HCL Technologies, Infosys, Mahindra Satyam, Tata Consultancy Services, Wipro

Technologies, ITC Infotech India Ltd and MphasiS among many others.

Like much of Southern India, Bangalore has a very high level of literacy and English is generally spoken. At the same time, India's growing population means that there is a ready supply of labour which holds back the growth of wages. Reflecting this, Bangalore remains the one of the world's best sources of relatively cheap, skilled English-speaking labour. Meanwhile, the success of the recipe has encouraged the state government of Karnataka to persist with it, encouraging business startups and inward investment. This is in stark contrast to many other parts of India. As a result, 40% of India's IT industry is based in Bangalore[20] with more than 500 companies generating US$17 billion in sales.[21] This success is continuing – two thirds of the new jobs created in Bangalore in Q2, 2014 were in IT[22] and nearly half the IT startups in India take place there. Some of the recent successful startups include online grocery store ZopNow, gaming company MadRat games and MagnetWorks Engineering, which provides real time dashboards, custom reports and analytic tools to small- and medium-sized businesses (SMBs).

Mukund Mohan, CEO of Microsoft Ventures in India, claims the city has a good "grassroots networking organisations for startups" and a "dynamic, consistent and ever learning startup system" that encourages entrepreneurship.[23] It also benefits from the fact that many of its people worked extensively in Silicon Valley before returning to apply their skills locally. Bangalore has many links with Silicon Valley, with companies such as Visa locating R&D hubs in the city. The city certainly has been booming, becoming sufficiently successful to be criticised by sources as diverse as *The Guardian* newspaper and the World Bank![24] Of course it is true that economic success in developing economies tends to generate a dangerously rapid pace of urbanisation, with an increase in squalid living conditions. Sadly these are the normal symptoms of the growing pains associated with economic success in emerging economies (and sometimes not so-emerging economies – there is similar pressure on housing in London though the results are less unpleasant). In an ideal world they would not happen but it would seem perverse to attack the economic growth that has done so much to reduce poverty in the emerging economies[25] just because some of its side effects are unpleasant.

India is now trying to create other technology hubs and will no doubt succeed over time with such a large and growing population. Yet Bangalore is well established as a cluster and with Indian growth likely to be a major element of world development in the coming years, its position is likely to be secure.

Beijing

The rise of Beijing as a tech and startup hub is in line with the Chinese government's plan to adjust its economic growth away from low-skilled labour-intensive export driven growth towards higher value industries such as technology. One of the factors driving this is demographics. The size of the Chinese labour force is already falling as a delayed result of the 'one child per family' policy and as labour becomes increasingly scarce, economic development needs to adjust from labour-intensive activities to higher productivity activities.

As with most things in China, the startup boom came both rapidly and on a very large scale. In March 2013, the local government of Haidian district in Beijing attempted to remake the formerly named Haidian Book City, a quiet street containing numerous bookshops, into Z-innoway, short for Zhongguancun Innovation Way. Since then 450,000 square feet of real estate has been reclaimed and the government spent millions recovering 250,000 feet more.[26] At the moment the street contains 339 startups, and is the focal point of a Silicon Valley-esque subculture in northeast Beijing.

In some ways China's tech industry has already created global giants. Firms such as Huawei (allegedly spun off from the communications arm of the People's Liberation Army and now an international communications giant) and Lenovo (who bought IBM's PC business) already compete globally at the high end of computing and mobile devices. Xiaomi already leads the pack in terms of market share in the largest smartphone market in the world; despite not being widely known outside China and Asia, it is the third largest smartphone manufacturer after Samsung and Apple.

When Alibaba (a Chinese e-commerce company) floated on the New York Stock Exchange with a record IPO (Initial Public Offering) of $25 billion – Facebook's IPO in 2012 was $16 billion in total – its owner Jack Ma became the richest man in China. The popularity of its IPO was partly spurred by the prospect of tapping into the Chinese consumer market with a population of

1.4 billion people. Alibaba, which controls a near monopoly at 80% of China's online shopping market, had an estimated market capitalisation value of $215 billion at launch. In that sense it was the fourth biggest tech firm in the world, behind only Apple, Google and Microsoft – such is the scale of the Chinese consumer base. Despite its fall in value as the initial hyper-enthusiasm waned, its value on 29th Jan 2016 remained as high as $168 billion.

Yet whilst firms such as Alibaba and Huawei were founded in special economic zones such as Hangzhou and Shenzhen, Z-innoway subsumes a startup culture on a smaller scale that is more reminiscent of Silicon Roundabout in London and Silicon Valley. A key characteristic of the district is 'startup cafes' such as Garage Café and 3W Coffee, which host startups meetings and 'accelerator' programs in a setting not unlike the coffee shops in downtown Palo Alto. Furthermore, the Chinese government is setting up a service centre that will reduce the time taken for a technology startup company to receive the necessary licenses and approvals, to four days from fifteen.[27]

The outlook for technical development in China has to be promising – though subject to any risks of social upheaval. Growth is likely to remain relatively strong, if more volatile than hitherto and as the economy moves up market, the demand for technology will grow. Yet one wonders whether the government's desire to maintain stability with controls on social media will conflict with the sector's development. Equally, traditional cultural values such as respect for authority might inhibit the freethinking and creativity that are associated with the growth of a creative technological hub.

Perhaps the most likely result is that the Chinese digital sector will continue to develop rapidly but in a lopsided way, with less focus on social media, on creativity and on wilder ideas but more focus on the industrial and technological side of the sector.

Hong Kong

Although they are not formally included in the MIT study, I have also looked at Hong Kong and Singapore as they are (or in Hong Kong's case were until recently) city states with some similarities to London. Like other governments, the Hong Kong government has invested heavily to develop a technology hub called Cyberport which had already virtually filled up as early as 2010.[28]

Success in Hong Kong partly reflected the emergence of mobile phone apps as a key area in software development. Mobile phones have had very high penetration in Hong Kong since Hong Kongers are, for the most part, early adopters of new mobile technology.[29] Mobile apps companies consisting of small teams of two to three persons using free development kits from Apple and Google can start easily, often with minimal external funding.

The increasing emphasis on mobile development means that even though wages and costs in Hong Kong are higher than in mainland China, they are affordable for entrepreneurs. Break-even points have fallen and many smaller firms don't even need venture capital financing; they can cover their costs based solely on mobile app sales. A small company can sell its apps to the world through the

Apple and Google app stores, and collect payment without restriction.

Local success stories include:

- Advanced Card Systems, a leader in smart card technology, ranked by Frost and Sullivan as one of the world's top three companies in its field, produce smart cards and smart card readers. In 2013 it had 289 employees and annual sales of $25 million.
- Outblaze has had a high profile since its founding in 1999, but as it is not quoted yet on any stock exchange many details about it are unclear. It produces apps and games including Pretty Pet Salon. It sold its messaging service to IBM for an undisclosed amount.
- PCCW is a significant technology and communications company. In 2000 it took over Cable and Wireless's assets in Hong Kong. It employs nearly 20,000 employees around the world and is quoted on the Hong Kong Stock Exchange and (in ADR[30] form) on the OTC (Over-the-Counter) Pink Market in the US.

Hong Kong's high population density makes universal wireless and high speed broadband widespread, making US broadband seem painfully slow in comparison. Some of Hong Kong's success reflects its access to the Chinese market but in a less regulated environment. For example, setting up a business in Hong Kong is much simpler and more straightforward than mainland China – a new company can be registered and bank accounts opened in two days, compared with an average 30–60 days in China.

Other growth areas for Hong Kong developers are

Facebook gaming and apps. Facebook is very popular in Hong Kong but is blocked in mainland China. Unlike in mainland China, websites are not censored by the Chinese government's Golden Shield, referred to by critics as the Great Firewall of China, or '#gfw'. Twitter and Facebook are freely accessible in Hong Kong, without the need for proxy servers as they are in mainland China.

In mainland China, the only three mobile operators are China Mobile, China Unicom and China Telecom. They all report directly to China's State Council, China's cabinet, and obtain the approval there for major business decisions. This applies even to the introduction of new mobile services, making them political, not business decisions. In contrast, Hong Kong mobile operators are unregulated and can currently introduce new services for consumers without the need for this political input. While mainland Chinese operators are held back waiting for decisions, Hong Kong operators just charge ahead in the competition for consumers. Hong Kong also has a very large creative sector based on its long established film industry. So far there has been relatively little spill-over into a domestic digital industry but based on what has happened in other centres this can only be a matter of time.

Hong Kong also has the great advantage of access to the Chinese market with many fewer regulations under the 'One Country, Two Systems' policy that is scheduled to last until 2047. It also has a highly educated population and – as demonstrated by the extent of the protests starting in the Autumn of 2014 – a rather more freethinking population than Confucian traditions might lead one to expect. How successful Hong Kong will be in the digital economy will depend on the attitudes of the Chinese government and on whether the freethinkers shift their focus from politics to business. My current forecast it that it will grow more slowly than London for at least the next four years but may accelerate thereafter.

Singapore

According to the Singapore government,[31] Singapore is Asia's most network-ready country, and is an "ideal location for international media companies to establish cutting-edge digital media services". The government claims that the interactive and digital media industry in Singapore "has grown by more than 1.5 times since 2008 to exceed S$2 billion in revenue today".

Singapore is widely recognised as having good intellectual property protection by international standards which have encouraged its emergence as an innovation hub. The Global Innovation Index[32] ranks Singapore as the seventh best in the world in 2014. To develop the country's capabilities in the area, the Singaporean government has committed US$1.3 billion to developing the country's R&D capabilities, enterprise innovation and entrepreneurship through the Research, Innovation and Enterprise Plan (RIE 2015). The Singaporean government is very aware of the need to build on the economy's spectacular success by taking its innovative and creative sectors "to the next level".

Committed to research and development Singapore has a range of research centres, such as A*Star's Fusionopolis and Biopolis, the National Research Foundation (NRF) Create (Campus for Research Excellence and Technological Enterprise), the CleanTech Park, the Tuas Biomedical Park, and Singapore Science Park. Government data points out that these are "home to over 1,000 research collaborations with leading players such as Hewlett-Packard, Fujitsu, BASF, Mitsui Chemicals and Nitto Denko".[33] Multinational corporations like Procter & Gamble and Danone have also established knowledge-intensive regional R&D centres in Singapore to develop Asia-specific products.

Yet it appears that although Singapore is one of the world's successful R&D hubs for industrial R&D, the country has not yet developed the coffee swilling, freewheeling tech culture that characterises areas like the East End of London. Perhaps this is only a matter of time.

The UK

The international analysis highlights a range of factors affecting the ability of other cities desiring to replicate the Flat White Economy in their own locales.

None of the centres have emerged without some involvement from the public sector at either the central or local level. Yet those sectors that are driven most by the public sector (France and Russia) seem least successful, particularly when measured in relation to the scale of public investment, though even those hubs are likely to be successful eventually.

Availability of skills is critical. Sometimes this emerges through proximity to centres of education, sometimes through proximity to cheap accommodation, preferably both.

In the case of London, the availability of skills seems to have benefitted particularly from immigration into the UK (see Chapter 7) and especially from the weakness of the Eurozone economy and the very high levels of youth unemployment in Southern Europe that have resulted from the Eurozone crisis. Indeed, London's Flat White Economy development has occurred despite relatively expensive accommodation (though rents have been driven up by the very scale of the immigration that has been associated with the Flat White Economy). Of course the analysis in this book shows that it is not purely economic factors that have stimulated the Flat White Economy. People come to London as much for the fun factor as for work and a liberal social environment (It is no coincidence that Bangalore, India's software capital, also has more pubs per capita[34] than any other city in India!). Fun is an important ingredient of the Flat White mix, though like many other factors, there are sufficient counter examples to show that even in less liberal social environments, a Flat White Economy might take off if there are sufficient other positive factors.

A supportive regulatory environment is also essential. Protection of intellectual property is important and sufficient rule of law to prevent predatory behaviour is likewise critical. The nature of the Flat White Economy is that most businesses will fail, and some will perform adequately but the real attraction is the possibility of becoming the next Facebook, Alibaba, Amazon or Google with investment returns in the billions of per cent, making their owners into some of the richest people who have ever lived. This

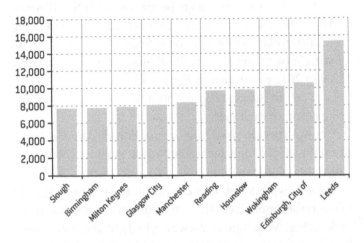

Figure 5.1: The biggest centres for the Flat White Economy in the UK outside London in 2012 (Source: Cebr)

also means that a relatively benign taxation regime (particularly for capital gains) is important.

We have applied the analysis above in various Cebr studies that have looked at parts of the UK that appear ripe to become new Flat White Economy hubs.

From analysis of the international drivers of tech hubs, key ingredients are: plentiful skills, good superfast broadband, availability of property, communal space for networking and a culture favourable to entrepreneurship.

We looked first at those areas that have existing jobs in the area within the UK. These comprise four hubs in the M4 Corridor area of Slough, Reading, Hounslow and Wokingham; five major urban areas which are Edinburgh, Glasgow, Birmingham, Manchester and Leeds and one other area less easy to categorise which is Milton Keynes. Of these, Leeds has by far the greatest number of existing jobs in the Flat White Economy Sector. The numbers of existing jobs in the Flat White Economy in each of these areas in 2012 are shown in Figure 5.1.

Each of these areas is of interest – we give a short description of each below.

The M4 corridor has long been a hub for the UK bases of major global high-tech companies, such as Intel, Microsoft and Vodafone.

The area is less focused on education and research, instead being rather more a centre of telecommunications (O2 are in Reading, Vodafone in Newbury) whilst there are world beating microchip manufacturers in Bristol and Bath.

Birmingham is second to London (though not a very close second) in business startups with 16,000 in 2013 – approximately the same number as in the London postcode EC1V. Birmingham Science Park Aston is home to the growing Innovation Birmingham complex. According to *The Guardian*[35], it currently houses 86 technology companies working on everything from games to business information services. Many of the companies are enrolled on the Entrepreneurs for the Future (E4F) programme, which combines free office spaces and communications with active support and a mentoring programme. The community gathers at meet-ups such as the monthly "Tech Wednesday" or the near-daily gatherings promoted by Silicon Canal, a not-for-profit group organised by, "tech people who run technology companies" whose ambition is to build a "tech ecosystem around ourselves and our companies." Innovation Birmingham Complex has plans to expand its premises by 120,000 square feet in the near future.

Manchester Science Parks (MSP) boast of five locations and 150 science and tech companies across the Greater Manchester region including the 'Corridor' centred around Oxford Road, traditionally the university sector in the city. In addition the £950 million MediaCityUK development in Salford was kick-started by the move of some BBC personnel and functions to Salford.

Cebr's report for the Glasgow Chambers of Commerce on the Glasgow media cluster[36] showed that a third of all the Scottish jobs in media were based in Glasgow with over 100 production companies and 300 facilities operators, whose combined turnover is about £1.2 billion annually. The three major national broadcasters – BBC Scotland, the Scottish Media Group (STV) and Channel 4 – are based in the Glasgow area, which is also home to a number of smaller, independent film companies and print and radio operations. These major media businesses are supported by a dense network of smaller TV production companies, advertisers, printers and other related businesses, adding to the richness and strength of Glasgow's media sector.

These facts are indicative of a media 'cluster' in Glasgow. This clustering (or gathering) of media businesses located in and around Glasgow has no doubt played a vital role in contributing to the sector's growth and development. Businesses in the same sector benefit from their proximity to other media businesses – this proximity facilitating knowledge sharing, business links and networking opportunities.

Clustering in Glasgow has mainly taken place in the

redevelopment and regeneration of the Clyde area known as 'Creative Clyde', which is home to the city's Digital Media Quarter and The Pacific Quay development, as well as incubator hubs such as Film City and the Creative Clyde Business Incubation Centre. Media businesses locating in this area no doubt benefit from their close proximity to organisations such as BBC Scotland, STV, Capital Radio, BIP Solutions, SECC, Film City Glasgow and the Glasgow School of Art's Digital Design Studio, which are all based there.

Skyscanner, a flight comparison website with 150 employees and a sister office in Singapore – relocated to Edinburgh in 2003. Edinburgh's startups are concentrated on the TechCube and many link up with Edinburgh Research and Innovation, a spin-off from Edinburgh University. *The Financial Times* in 2013[37] reported 244 business "spinouts" from Edinburgh University since 2001 – a high percentage of the UK total. Prominent in Edinburgh's digital cohort is FanDuel, a leader in one-day fantasy sports games reportedly a 'billion dollar' company in the making with investment from and market opportunities in the USA.[38]

The Yorkshire Post has teamed up with Leeds Beckett University (formerly Leeds Metropolitan University) to help startup digital operations achieve their growth potential in their city. Together they are setting up the Leeds Digital Hub which will offer city centre office space, mentoring and training to the companies as well as encouraging collaboration. Plans for the New Dock area in Leeds (with one eye to the anticipated High Speed 2 (HS2) rail link intend to shape it as a Northern 'Silicon Roundabout'.

Perhaps the most surprising area on our list is Milton Keynes. It neither fits the M4 corridor model nor the major city model. But the Centre for Cities as reported in *The Financial Times* has pointed to Milton Keynes and Northampton as the fastest improving economic environments in the UK.[39] The arrival of Network Rail in 2013, which rationalised a range of offices spread around the UK to centralise 3,000 jobs in Milton Keynes, has both driven population growth and stimulated supply businesses. New businesses in the area range from AirWatch, a global mobile device management company with 150 employees to the world famous Red Bull motor racing team. These condensed pictures show that London is not the only centre for the Flat White Economy in the UK.

Cebr's approach to predicting areas of growth has been developed through the construction of an innovation index for Johnson Press plc looking at different areas of the UK at a subregional level so as to understand where the innovation potential in the UK is at its highest[40.] This has given us an indication of potential. The index takes account of skills, innovation success and track record and exposure to high technology in products. It concludes that while many of the potential areas for development of the Flat White Economy are likely to be in the South East region, the East of England and the South West of England, there are also likely areas with potential for success in Manchester and Cheshire, in Yorkshire, in Derbyshire and in the East of Scotland.

Taking both together, there is clearly scope for a range of spin-off technology clusters in the South, East and West of England but clear potential in Manchester, Leeds and

Edinburgh. From personal experience I would also expect Glasgow to offer such a cluster based on its already strong media presence. While the plethora of music and cultural developments in Brighton are also likely to spill over into digital growth.

But as East London's digital economy matures, it is passing the beacon for fastest growth to South London. In many ways South London (and particularly South East London) is the new East London.

What all this indicates is that the London Flat White Economy model does not have to be the only one in the UK. Clearly the highest energy of high technology activity is likely to be in the area where connections are made and where finance is available virtually immediately, as in London. However, the technology sector contains many different types of business and even within the same business sector, there is a lifecycle with changing requirements that can vary over time.

This may prove to be particularly the case as parts of the technology sector mature. As that occurs there is likely to be an increasing requirement for a more systematic and disciplined approach, which might be more easily achieved outside the buzz of London. There is scope for a range of types of high-tech businesses and one of the UK's advantages in this respect is the range of different locations available.

CHAPTER 6

The UKs Unbalanced Economy

One of the themes of this book is that the London economy is doing well – growing much faster than the equivalent cities in Asia (Hong Kong and Singapore) or North America (New York). However the UK economy is – although one of the stronger economies in the Western world – not doing *that* well and if London's stellar performance is removed, the performance of the rest of the UK looks decidedly lacklustre, only appearing good in comparison with European economies held back by a range of problems including the Eurozone crisis.

My concern in this chapter is about the unbalanced nature of UK economic growth.

My argument is that although in an ideal world growth would be spread evenly – like Marmite – in the real world this just does not happen. I argue that the position of the rest of the UK outside London and to some extent the South East would be very much worse if London were not so successful. This is because of the scale of the beneficial-spillover effects to the other parts of the UK economy from London's growth.

I then go on to list some ways in which we could improve on the current situation by restructuring regional policy in the UK.

What makes cities succeed?

It is worth starting with some observations about what makes cities succeed and fail.

The acknowledged expert on the subject is Harvard academic Edward Glaeser, author of *The Triumph of the City*.[1] He and I have been working along parallel lines on the driving forces behind the economic growth of cities for more than twenty years now and although we have never corresponded, our views appear to have moved in similar directions.

Glaeser's key insight is that cities depend on transport. Historically it has been difficult for land transport to feed large cities which has limited the growth of those cities who do not have access to sea or river transportation. He points out that before vehicles using anything other than animal power, the cost of land transport was thirty times higher than that of transport on water. And I am indebted to my predecessor but one as Gresham Professor of Commerce, Michael Mainelli, for pointing out that London in the Victorian era was largely fed by barges. The report from the Joint Select Committee of the House of Lords and the House of Commons on the Port of London Bill in 1903[2] quoted a Mr Erskine Pollock who claimed that 76% of London's imports at that time arrived along the Thames. Mainelli also argues[3] that as a junior researcher helping a Harvard Lab for Computer Graphics and Spatial Analysis project on cities in the 1970s, "two of us spotted this 'seizure' somewhere about 500,000 to 750,000 people and oxcarts depending on your parameters". Beyond this point cities with only land transportation (pre-motorised transport) seized up.

London's expansion to the largest city in the world predated the age of the railways – Mainelli argues that this was because of the ability to supply the city by the Thames. But the growth of cities in recent times has been facilitated by motorised transport which has enabled large cities to be fed without generating unsustainable congestion. As a result, we have seen the emergence of a wide range of mega cities around the world, especially in large emerging economies.

During the 1970s there seemed to be evidence in the developed economies of a slowing down in the growth of cities. But since then, many cities have had renewed expansion. Glaeser relates this to two key factors: technology and globalisation. He argues that these have both placed a premium on 'smartness' and that your smartness depends on your interactions. He points out that there is a disconnect – in less skilled cities there is little correlation between city size and productivity. But in 'smart cities' as much as 45% of productivity is a function of the city size as the increased connections that can be made increase the creativity and economic potential of the employees.

The underlying thinking is similar to that of the economics of agglomeration and of clustering. In a modern city – and this is increasingly the case – intelligence based work is improved in its quality of thinking, which is partly a function of connections, by increased numbers of interactions. And increased numbers of encounters increases the number of connections and hence creativity. Arguably this is why the London economy has suddenly surged ahead again in the 21st century.

London's historical growth

London has had three periods of spectacular expansion and seems to be at the beginning of a fourth. The first was in the Roman era. The estimated population grew from fewer than 100 people estimated in 50 AD to about 50,000 (just over 3% of England's population)[4] in 140 AD.

After the Romans left, the population of London fell back both absolutely and relative to the rest of England and it was only with the consolidation of government in the Tudor period that London's share of the UK population started to grow again. From 1540 to 1700 London's population rose from 120,000 to just short of 600,000 – from 4% of England's population to 11%.

London's third and greatest period of expansion was during the 19th century when the population grew from just below a million in the census of 1801 to 6.5 million in the 1901 census. As a share of England's population, London grew from 12.4% to 21.7% over this period. From 1831 to 1925 London was the world's largest city by population.[5]

London's actual population peaked pre-World War Two at just below 9 million. It then fell to a low point of 6.8 million in 1981, before starting to recover with the development of the financial service economy of the 1980s and the business service and headquarters economies of more recent years. London's population in 2014 is now roughly back to its 1939 peak and is projected in the latest London plan to grow to 11.3 million by 2050.[6]

There is very little historic data for London's GDP but the economic historian Nick Crafts[7] has generated a series based on the earlier Geary Stark Regional data but adjusted to take account of tax data. His estimate for London's share of England's GDP in 1911 is, at 27.7%, actually higher than any current estimate for 2014. Most estimates suggest that London's share at present is around a quarter.

What is interesting is that many commentators who think that London's economy is excessively dominant as a share of the UK or England, do not realise that the share is in fact slightly below previous peak levels. For example, a Bloomberg article by Tyler Durden[8] claims that 'London's contribution to UK GDP is at record high' though since the article includes no historical data it is not clear how that claim is meant to be substantiated. The historical data we have examined here suggests that the current share is slightly lower than its peak pre-WW1.

Relative importance of cities elsewhere in the world

There also appears to be a belief that London's share of the UK or English economies is uniquely high by the standards of major cities in other countries. Danny Dorling in the *New Statesman*[9] claims that London divides the UK in a way that no other country in Europe is divided.

The major city that dominates its economy the most is Dhaka, the capital of Bangladesh. This counts for 60% of Bangladesh's economy. Bangladesh is a poor country, with a per capita gross domestic product of $1,400, but it is the world's seventh-most populous country. Cairo is dominant in Egypt's economy, contributing about 50% of GDP, while Mexico City, with 19.3 million inhabitants, generates 40% of Mexico's GDP. Other economies with dominant leading cities include France (Paris, 30% of GDP), South

Africa (Johannesburg, 36%) Argentina (Buenos Aires, 46%), Austria (Vienna, 36%) Sweden (Stockholm, 32%) Japan (Tokyo, 33%) and Korea (Seoul, 23%).

Dorling suggests that one of the problems with London is that although its share of GDP is by no means as large as the shares for some other major cities, "London is small (in population terms) compared to say Seoul or Tokyo, consequently its economic dominance is even more pronounced". He appears to mean that the share of GDP is relatively high compared with the share of population. This partly reflects London's relatively high share of commuters (which partly in turn reflects the fact that the definition of London for these comparisons is a relatively tight one whereas some of the other cities with which comparison is made are defined to include a much larger area that in effect covers many of those residents who in London are treated as commuters from outside). But it also reflects the fact that London's economy works effectively to make people who work in London highly productive. This should hardly be a cause of criticism – surely the fact that London helps people be more effective and economically successful than they would be elsewhere should be a cause for praise, rather than criticism.

Some reasons why inequality is a bad thing and some reasons why it is not

In theory whether London's success is good or bad for the rest of the UK should be a factual matter that can be disentangled by careful analysis and modelling. Like all attempts to model the counterfactual there cannot be complete certainly about the conclusions but it ought to be possible to narrow the scope of disagreement to a small range. But in fact much of the argument about whether London's success is bad for the UK seems to have taken place in the absence of key facts. Moreover there seems to be a link between those who think that there is a problem with London's success and political orientation.

One reason for this is that some of those concerned with the gap between London and the rest of the UK are not especially concerned with whether the London economy is beneficial for the rest of the country as such, but are simply concerned about the inequality between the two. This is out of line with the teachings of traditional welfare economics which have adopted the Pareto principle. The Pareto principle is that a change that makes no one worse off and some people better off is a change for the better (a Pareto efficient improvement). But if inequality is the concern, so-called 'Pareto-efficient' changes may or may not be an improvement depending on the initial positions of those made better off.[10] Many modern commentators seem to think that inequality in itself is the evil to be avoided, regardless of its cause[11] despite the inconsistency of this approach with a fairly long tradition of welfare economics.

One reason why some of the modern theorists of inequality think that it is a problem in itself is the concept of relative poverty. The thinking is best expressed by the interview which Natalie Bennett, the leader of the UK's Green Party gave with *The Economist* where she is alleged to have said: "to be poor in India wasn't so bad as to be on benefits in Britain because at least everyone else there is poor too".[12] She has been criticised heavily for her comment. *The Economist* claimed "That is contemptibly naive and also a shame".[13] Nigel Farage leader of UKIP and never one to be short of a pithy comment responded in a tweet "What utter drivel, highlighting a major lack of critical thinking and compassion".[14]

There is a second way in which inequality can be a problem in itself – if the economy operates such that either production or consumption is a zero-sum game. On the production side, the issue is whether those who become wealthy do so at the expense of others. Here there seems to be a correlation between concern about inequality in and of itself with those countries where there has been a feudal history – for example most of Europe – and conversely those newer societies which do not have a feudal history.

In a feudal society, land is the scarce resource and is in limited supply. Moreover it was often either allocated by the accident of hereditary good fortune or by force. Hence inequality typically meant that someone had more at someone else's expense. The generation of wealth by one person was generally at the expense of some other person.

But in the modern information era and especially in the Flat White Economy, most of the wealth is freshly created and a surprisingly small share is inherited. The reason that the wealth exists is because the person who owns it has created it in some way. There is scope to argue about whether all the ways in which people create wealth in the modern world are fair – some areas of monopoly and apparent extortion quite clearly are not. But it is unlikely that the wealth would have been created on the scale that it has been, had it been rational to expect that it would be confiscated. There is fairly good statistical information on the inverse relationship between wealth creation and the extent to which that wealth is taken by the government in taxation.[15] So in general it is reasonable to suppose that in an information society much of the wealth created is additional rather than being taken from someone else as in a feudal society. In these circumstances redistribution to reduce inequality alone is likely to destroy a proportion of the wealth that is redistributed. It is not a zero-sum game.

Because of the different historical sources of wealth, public opinion in relatively new economies like the US, Hong Kong and Singapore that have not experienced a feudal system appears much less concerned about inequality than opinion in countries with a feudal history like much of Europe, where I suspect that because of the feudal origins of society, wealth is considered highly suspect by a significant proportion of the population. The tragedy is that in Europe, political and social views are being based on an historic view of society that is of diminishing relevance in a modern information society.

But while the creation of wealth in the modern world is normally of a kind where much of the wealth creation is not created at the expense of others, there is still an extent to which the prosperity of the wealthy worsens the position of others. This comes largely through the consumption effects for property. Property, especially in cities, is scarce for both natural reasons and because of self-imposed regulations such as planning laws. Because of this, greater wealth for some means higher property prices and hence a standard higher cost of living for others.[16] 16 In turn this means that the standard of living for those who have to pay higher rents is reduced. It is in this sense that those arguing that London's prosperity boosts inequality have a point and some of them have been demonstrating against gentrification to make that point apparent.[17] As I show later in the chapter, I would prefer to see the growth accommodated by more relaxed planning regulations helping to bring down housing costs in London rather than by eliminating the economic growth. The UK has all too few sources of economic growth for us to damage our most buoyant economy and our most innovative sector.

In the end, most people's moral calculus tries to generate a balance between the positive wealth generating effects of some people becoming wealthy and their spin-off benefits for the rest of society through taxation and other linkages against the negative impacts of the wealthy increasing the cost of property for others. Obviously, to the extent that the supply of property can be freed up, the negative impacts of inequality can be minimised while maintaining the positive benefits.

What are the linkages between the London economy and the rest of the UK?

So this chapter will not look at simply whether London contributes to increasing inequality. It clearly does and for those who think that in itself this is a bad thing, there is very little that can be said to contradict them. What the chapter does instead is apply a different test: does London's growth mean that the rest of the country is better or worse off? This I believe is the true test of whether

London's prosperity is good for the rest of the country.

To test this argument it is necessary to evaluate those links where London positively affects other parts of the UK economy and those were London negatively affects these other parts.

The positive links are:

1. Trade between the regions and London enriches both London and the other parts of the country.
2. Transfers from residents of regions outside London earning incomes in London which are spent in their own regions tend to benefit those regions outside London (There are also some flows in the opposite direction but they are much smaller).
3. Net fiscal transfers from London to the other regions – although public spending per capita is higher in London than elsewhere, this is dwarfed by the disproportionate tax yield in London which is spent in other regions that makes a huge contribution to other regions.
4. People and skills migration – this can work in both directions – London takes skilled people from other regions but also provides skilled and experienced people for other regions.

The negative links are:

1. London's success leads to a higher currency value than is competitive for the rest of the UK.
2. London's success leads to a tighter monetary policy (higher interest rates) than would be appropriate for the rest of the country.

3. Skills and talent being drained from other parts of the country to London (the flip side of point 4 above).
4. Loss of critical mass in other parts of the country as the talent pool is drained.

There are four published reports historically speaking that consider aspects of this that were prepared in the early part of the last decade:

(i) Canary Wharf: London's importance to the UK regions, (Cebr report for Canary Wharf Group, 2003)
(ii) Greater London Authority: Growing together: London and the UK economy, 2005
(iii) Corporation of London: London's linkages with the rest of the UK, 2005
(iv) GLA Economics Briefing Note 5: Has London continued to export taxes in 2003/04?

In addition there is a very detailed report by the South East Regional Assembly in 2005 which looks in much greater detail at the links between London and the rest of the South East of England.

There is also a very up-to-date study on fiscal transfers prepared by Cebr for the Corporation of London in November 2014.

a) London's Importance to the UK Regions

This was the first report on the subject[18], which has subsequently been followed up by other reports.

Its conclusion was that the other UK regions are

interlinked with the London economy and that far from London's economic growth being at the expense of that of other regions and countries, the other parts of the UK *benefit* from a successful London economy. This has been confirmed by the subsequent reports.

The report looks at four issues:

- The demand created in the regions by the goods and services that London buys from them.
- The benefits which accrue to the regions from that element of the earnings of London commuters which are spent locally.
- The regional allocation of the net fiscal contribution that London makes each year to the UK Exchequer.
- The loss that the UK economy would suffer were the dominant London financial sector to be scaled back to the scale of the European average.

The report is highly quantitative. It has two unique features:

- A quantified analysis of the impact of higher growth in London on economic trends in the regions.
- A detailed model-based analysis of what would be the impact on the regions if London's GDP growth were to be curbed, for example, through a less successful financial services sector.

It is the only report to quantify the extent to which the rest of the UK is influenced by interest rate policy set for the UK as a whole including London; which by implication means that a faster growing London is likely to push up interest rates for the whole of the UK and hence have some negative impacts on the other parts of the country.

Through careful quantification the report shows that the negative impacts of London on the rest of the UK are very much smaller than the positive impacts. The positive impacts are more than twice as great as the negative impacts.

b) Growing Together: London and the UK Economy 2005

Table 6.1: Correlation between economic growth in London and the rest of the UK, 1983–2004

Regions and countries of Great Britain	Correlation coefficient
South East	0.80
East England	0.81
South West	0.64
East Midlands	0.45
West Midlands	0.73
North West	0.73
Yorkshire and Humberside	0.56
North East	0.22
Wales	0.55
Scotland	0.27
Northern Ireland	0.36

Table 6.2: Percentage change in employment, 1989–2001

Regions and countries of Great Britain	% change
South East	23.7
South West	21.2
East of England	18.8
Scotland	17.4
London	15.3
East Midlands	12.5
Wales	11.7
West Midlands	10.8
Yorkshire and Humberside	10.2
North West	9.9
North East	7.2

This report looks at a range of links between the London economy and those of the rest of the UK.[19] It concludes that London's growth is not at the expense of the rest of the UK, but that London and other UK regions and countries are interdependent.

A unique feature of this analysis is research into the correlations between GVA growth in London and other regions and countries. This shows powerful correlations between annual economic growth in London and many of the other regions and positive correlations for all regions (however there are some statistical issues with the result which are not satisfactorily handled so the work should probably be examined carefully). The results are shown opposite.

The other elements in the proposed links are: commuting and migration (the results are very similar to the earlier Cebr report but there is no analysis of the impact of commuters' spending in their own regions); trading links (though there is no breakdown by region and the report is based on OEF and Experian data already published); London's role as a world city and London's specialisation.

c) London's Linkages With the Rest of the UK 2004

This report looks at a range of links between the London economy and those of the rest of the UK.[20] It concludes that "rather than rivalling other regions of the UK, London's success appears to have complemented and supported growth elsewhere in the economy". As with the other reports, this report looks at migration and commuting, including the impact of commuters' spending in each region, London as a source of skills, London's trading links within the UK, London's position as a specialist provider of a range of services within the UK and the issue of whether London's housing market has a negative effect on those of the rest of the UK. The report does not look at fiscal linkages, although it notes that London has a disproportionately low share of public sector employment.

d) Has London Continued to Export Taxes in 2003/04?[21]

This report uses official statistics for tax receipts allocated in a variety of ways and data for the breakdown of public expenditure to estimate total tax receipts from London and total public expenditure in London in 2003/04. The estimates suggest a net transfer from London of £1–7 billion. The calculations are reputable, if probably on the cautious side. What they fail to take into account, however, is the overall fiscal position of the UK government for the year. In 2003/4 the UK government in total borrowed £35.4 billion.

More conventional methodology for calculating the fiscal contribution of individual regions (for example the work for the World Bank of Prof R. Prud'homme)[22] normalises the overall tax/spending position for the government as a whole to work out what would be the regional imbalances if government borrowing was zero. That approach provides an estimate of regional transfers that is unaffected by the government's overall fiscal position. It is the approach used in the Cebr calculations.

(e) Symbiosis or Sibling Rivalry: The Future Links Between London and the South East[23]

This report looks in great detail at the links between the economies of London and the South East of England in the context of the London Plan and the proposed plan for the South East of England.

The report concludes that the regions of London and the South East are intimately linked:

- According to the 2001 Census, 371,000 South East residents worked in London in 2001 while 128,000 Londoners worked in the South East.

Report	Growing together	London's linkages	London's importance	Symbiosis or sibling rivalry	Does London export taxes?
Published by	GLA economics 2005	Corporation of London 2005	Canary Wharf Group 2003	South East Regional Assembly 2005	GLA economics
Prepared by	GLA staff and Experian Business Strategies	Oxford Economic Forecasting (OEF)	cebr	cebr	Internal staff
Key conclusions	London's growth is not at the expense of the rest of the UK, but that London and other UK regions and countries are interdependent.	'rather than rivalling other regions of the UK, London's success appears to have complemented and supported growth elsewhere in the economy'	Far from London's economic growth being at the expense of that of other regions and countries, the other parts of the UK benefit from a successful London economy.	The regions of London and South East are intimately linked	London continues to subsidise the rest of the UK fiscally by an amount that – ignoring the deficit – is between £1billion and £7billion per annum
Unique features	Analysis of correlation of economic performance in different regions showing generally high levels of correlation.		Economic simulation of impact of London doing badly on other regions showing that gains from lower interest rates are less than half of job losses from reduced complementary employment.	The most detailed analysis of regional links with London, showing how the South East economy depends on London's success.	

Table 6.3: Comparison of different reports on impact of London on the rest of the UK.

- The report estimates that the gross incomes of commuters from the South East to London amounted to an estimated £22 billion in 2002 – equivalent to one sixth of South East gross value added.
- The report calculates that businesses in the South East 'exported' to London an estimated £18 billion worth of goods and services in 2002.
- The South East is the major destination for people moving out of the capital; a net 27,000 people moved 'from the smoke' in 2001 according to the 2001 Census.
- The report concludes that £15 billion of government tax revenues in 2002 accrued from the economic linkages between the regions and the report calculates that 15.5% of London's effective labour catchment is located within the South East.

All these analyses are fairly out of date today. But there has been no evidence that would contradict their unanimous opinion by different (even rival) teams of economists using alternative methodologies that London's success is on balance good for the rest of the UK.

Fiscal transfers

There is one area where the analysis of the links between London and the rest of the UK has been updated. This is the area of fiscal transfers, where the report produced by my colleagues at Cebr for the Corporation of London[24] in November 2014 has updated all the earlier data. This report not only looks at the current situation but also projects the situation forward to 2034/35.

The key conclusions of this report are:

- The UK economy is recovering from a tumultuous period following the deep recession of 2008/09, and is set to see the strongest growth in 2014 since before the financial crisis, at just under 3.0%. However, this rate is projected to slow over the medium term, constrained by tightening fiscal policy and normalising business and consumer confidence, falling back to just below 2% for the years to 2020.
- London has seen strong economic expansion, with estimated real term growth in the capital of 4.2% in 2014. Although this is projected to slow in the coming years, the rate is predicted to remain close to the 3.0% mark for the years to 2020.
- This faster expansion is due, in part, London's role as a financial and business centre, and its concentration of highly productive sectors such as financial and insurance services and professional services.
- As a result of London overall's buoyant economy, it plays a major role in generating tax revenues for the exchequer, at an estimated £127bn in 2013/14. This is roughly equivalent to the tax take from the smallest five regions and countries of the UK combined.
- London and the South East also provide the only significant net positive contribution to the public finances – at an estimated £34bn and £22bn respectively in 2013/4. Although these surpluses were smaller following the 2008/09 recession, these two regions have been the main contributors to public finances over the past ten years. By comparison, the UK public finances as a

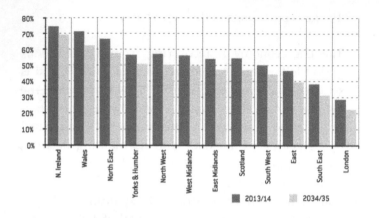

Figure 6.1: Public sector spending as a share of regional GVA (constant 2010 prices) 2013/14 and 2034/35.[25] Source: Cebr.

whole were in deficit by £98 billion.
- Austerity measures and changing economic situations have significantly reduced public expenditure on a per head basis in London; public sector current spending fell by 5.1% in London between the 2009/10 peak and 2012/13 in real terms. This compares to a 2.3% decline for the rest of the UK.
- London's public spending as a share of economic output is projected to fall from 30% in 2004/05 to just 23% in 2034/35 – indicating a very healthy private sector.

The data in Figure 6.1 shows that public spending in London is only 28.6% of GDP compared with figures in the rest of the UK ranging up to 74.4% of GDP in Northern Ireland.

Figure 6.2 shows the extent of the fiscal surpluses in London and the South East. Much of these surpluses were or

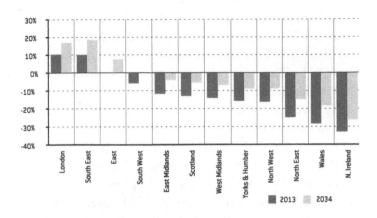

Figure 6.2: Regional surplus/deficit as share of local GVA, constant prices 2013/14 and 2034/35.[26] Source: Cebr.

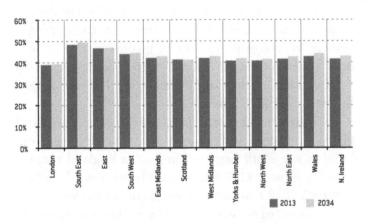

Figure 6.3: Tax revenues as a percentage of local GDP 2013/14 and 2034/35.[27] Source: Cebr.

would be actually earned in London since income taxes and National Insurance taxes on commuters are allocated to the region of residence rather than the region where the income is earned. The London surplus was £32 billion in 2013/14 and the South East surplus was £21 billion in that year. These £53 billion were allocated to fund deficits in *every other region* except the East of England (even that region would have been in deficit if the tax payments by commuters into London were allocated to the London economy!).

Figure 6.3 shows the tax receipts as percentages of GDP and shows in effect how commuting pushes up the tax receipts for the South East of England and the East of England (London receipts look low because commuters' tax payments are allocated to the region of residence, mainly, the South East).

The analysis for 2013/14 showing the extraordinary scale of fiscal transfer from London to the Rest of the UK is spectacular. In effect, once the impact of commuting is allowed for, roughly a fifth of all income earned in London is used to finance the rest of the country which by itself still ran a deficit of £90 billion in that year despite the transfer from London.

But projecting forward to 2034/35 on current fiscal plans, the scale mounts dramatically. The surpluses from London, the South East and the East of England are projected to grow to £170 billion. At this point nearly 30% of London's GDP represents an excess of tax payments over public spending once commuters are taken into account.

This massive scale of fiscal transfer reflects the UK's highly progressive (and increasingly so) tax system.

The net impact of London

From the analysis above, the net impact of London on the rest of the UK is substantially beneficial.

This of course is a static analysis, though it shows benefits outweighing the costs on a huge scale.

Modelling dynamic systems is fraught with dangers. Perhaps the best known such models are those attempting to model climate change which have had surprisingly little success in predicting climate changes, despite the fact that there is a large scientific consensus agreeing with the principles embodied in such models.

I have not attempted to create such a dynamic model of the relationship between London and the rest of the UK and so this remains a potential weakness in the argument that London's success is on balance a benefit for the rest of the UK.

Nevertheless, the conclusion is a strong one, even if not incontrovertible.

How to improve the relationship between London and the rest of the UK

Rather than debating whether London is on balance good for the rest of the UK (which I have argued *is* the case), it is more useful to look at how best to boost the benefits from London's wealth and minimise the disbenefits.

What is clear is that the cost of living outside London is very much lower than in London and the weakness of the non-London economy means that there is substantial availability of skills and other resources.

The trick is to make the London and non-London economies increasingly complementary.

What happens at present is that taxes raised on economic activity in London are largely used to finance high levels of public spending in regions outside London.

While some of this public spending (particularly on infrastructure and education) probably gives a net boost to enterprise, much of it does not, even if it adds to the quality of life more generally. But one of the effects of providing heavily funded public spending in the economically weaker countries and regions of the UK is that it tends to divert the energies of talented young people towards the public sector rather than the enterprise sector. Ultimately this often means that their talents are biased to consuming resources rather than creating them.

This in turn seems to be reflected in reduced entrepreneurship in the weaker regions. For example, in 2013 there were 1,266 businesses per 10,000 residents in London and only half this number (exactly 633) in the North East of England.[28]

My most controversial proposal is to change the relationship between tax and spending. Instead of the taxes raised in London being used to support high levels of public spending in the regions, they should be used instead to reduce taxes on income and corporate taxes for economic activity being generated in the regions.

This huge amount of money being transferred from London to the rest of the UK should be used to make the rest of the UK more competitive.

The 'Barnett formula' which tries to guarantee spending levels should be replaced by an incentive formula based on the real level of income inequality (i.e. after adjusting for the cost of living) between the regions so that weaker regions would have lower taxes on both personal and corporate incomes.

This will turn the economically weaker parts of the UK into tax havens.

It is pretty obvious that this is likely to help rebalance the UK economy – indeed it could do so remarkably quickly. Moreover, since the policy is likely to lead to better utilisation of resources, it is likely to lead to faster economic growth overall and to permit both tax reductions and ultimately better public services as well for the whole country.

Whether it would prove politically acceptable depends on whether the relevant vested interests are allowed to veto a proposal that has considerable economic advantages.

It is not quite clear whether it would be consistent with EU rules. These only allow regional aid and subsidies to businesses under fairly strict conditions. But on the other hand, EU rules do not actually force countries to have the same rates of tax in every region. It should not be beyond human abilities to find a way of constructing the system for varying tax rates by region to keep it within EU rules.

This proposal as it stands has the potential to make a huge change to the economic shape of the UK by turning regional disadvantage into regional incentives.

But this isn't the only thing that could be done to help reverse the trend for regional gaps to widen. The second proposal for diminishing regional imbalances within the UK is to turn the brain drain to London into a skills drain out of London.

It works like this. Clearly many bright people from around the country are tempted by London's shining lights and lively economy. London isn't just successful, it is also a lot of fun, with the only real negatives being a fast pace of life and expensive property.

But people do not always want to spend their whole lives in London's fast lane. The trick is for those who move from a region or country of the UK to London on a probably temporary basis to have an easy journey back.

There need to be skills agencies in the regions to attract back those from each region who have migrated to London, when the time is right for them to move back. When these people have developed a career and built their connections, they are then able to keep these connections even when they move back to their home regions. Because of the high cost of property in London, one of its disadvantages is that it does not provide as comfortable an environment as other places for bringing up children. Many who reach the age of having children would like to return to their home regions if they were able to preserve their economic advantages.

So people who have migrated from the regions and have developed their careers in London should be encouraged to set up businesses back in their home regions, taking advantage of the skills and networks that they have developed in London and taking advantage of the lower costs and higher standards of living in their home regions. In this way, London and the regions can develop interactive relationships that benefit both.

The incentives to develop such relationships do not necessarily have to be financial. The financial advantages implied by the proposal to replace high public spending in the regions with tax advantages will probably be sufficient as financial incentives. But the trick to encourage entrepreneurs to keep their links with their home towns and communities is probably mainly social – each region should compete to make its migrants feel that they will be given significant social status when they return. One way of promoting such competition might be to publish statistics on comparative success in attracting returners! In that way, parts of the UK like Scotland, where the local elite have been traditionally hostile to returners, might be named and shamed when it can be seen how much this cliquishness is costing the local economy.

These are radical proposals, particularly the proposal to replace heavy public spending outside London by turning these regions into tax havens. But my forty years of experience as a professional economist make me believe that they would be by far the best ways to reduce the degree of inequality throughout the UK and help revive the UK regions. If politicians are serious about wanting to reduce the huge scale of inequality between regions in the UK, I challenge them to take on the vested interests and implement these proposals.

CHAPTER 7

Immigration – the UK's
Secret Economic Weapon

London has a long history of immigration since Roman times. In the 13th century London was described as "overflowing with Italians, Spanish, Provencals and Poitevins"[1] in an uncanny precursor of the situation in the early 2010s following the Southern European economic crisis. This chapter looks at the way in which the latest phase of immigration has affected the London economy and especially the Flat White Economy.

It argues that the unique combination of circumstances resulting from a requirement for a new highly creative labour force and the widespread availability of skills from a combination of the indigenous population (from many parts of the UK) and migrants – especially from the economic crisis hit areas of Southern Europe – has created an unusual mix which has been a powerful driving force behind the growth of the Flat White Economy in London.

The history of immigration into London

London is generally considered to have been founded by the Romans around 50 AD. When founded, the bulk of the population by definition would have migrated to the city. When London was reinvented as an Anglo Saxon City around 600 AD the majority of the population were migrants from elsewhere in the UK. But London's central position and entrepot trade made it attractive to those from elsewhere. The name of Lombard Street in the City of London pays homage to the bankers who emerged there from Northern Italy.

According to a study by the London School of Economics[2], "By the 15th century there were estimated to be 1850 foreigners in London, rising to 3,000 in 1501".[3] Further waves of migrants included Huguenots fleeing Catholic persecution in France and Jews of different origins also often fleeing persecution after they were allowed into the country by Oliver Cromwell. In addition there was apparently a population of between five and ten thousand black people in London by Georgian times.[4] These waves of migration were associated with economic growth, rising wages and costs and an energisation of the economy according to the LSE study.

Further migration in the 19th century was driven by both demand-side and supply-side factors. Pogroms in Europe led to increased Jewish migration, while famine in Ireland led to the emergence of an Irish population of over 120,000 in the city by the end of the century. More recently in the 20th and 21st centuries migration has been driven by a range of factors as the cost of travel dropped sharply.

The wars of the first half of the 20th century promoted migration from persecuted groups and from occupied countries – though bombing during the Second World War made London less attractive and damaged the housing stock.

In the mid-20th century migration into London was

mainly from former Commonwealth countries though this slowed as increasing restrictions were introduced. In the latter years of the 20th century migration was driven by a range of factors which included asylum seeking; those taking advantage of the European Union rules for free movement of labour once the UK had joined the EU on 1 January 1973; and demand pull factors as the growing London economy started to look outside its boundaries for skills.

The LSE study shows a detailed table (Table 7.1) indicating how the migrant composition of the London labour force changed between 1986 and 2006 with the proportion of migrants from former British colonies dropping from 76% to 59%.

Table 7.1: The cosmopolitanisation of London's population 1986–2006

	1986	2006
Foreign born population	1.17 million	2.23 million
Proportion of total	17.6%	30.5%
Share from former British colonies	76%	59%
Main countries of origin	Ireland, India, Kenya, Jamaica, Cyprus, Bangladesh	Ireland, India, Kenya, Jamaica, Cyprus, Bangladesh, Nigeria, Poland, Sri Lanka, Ghana, South Africa, Pakistan, Somalia, USA, Turkey

The pace of migration quickened during the 21st century particularly after the financial crisis of 2007–8 and after the accession of the Eastern European countries into the EU. By the time of the 2011 census, only 44.9% of the

Table 7.2: Ethnic origin London's population 2011 Census

	Per cent of total
White British	44.9
Other white	14.9
Asian	18.4
Black	13.3
Arab	1.3
Mixed	5.0

Source: 2011 Census. Office of National Statistics (Note: these figures reflect self-identification).

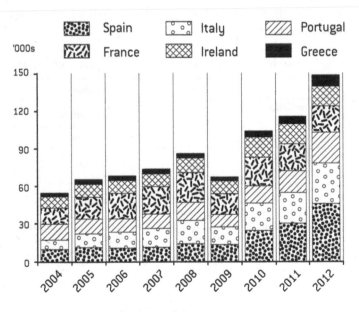

Figure 7.1: Southern European Migration into Britain since 2004. Source: Office of National Statistics, 1997, 1998.

population of London described themselves as white and British born.

This relatively more successful UK economy that emerged after the 2007 economic crisis and the knock-on effect of high unemployment (and particularly youth unemployment in Southern Europe and Ireland) has meant that a fresh wave of migration emerged, as is shown in the chart in Figure 7.1.

Politics

Partly reflecting these waves of immigration in recent years, the subject was a major political issue in the UK in the run up to the 2015 general election. Prime Minister David Cameron planned to have a quota for the number of National Insurance places (in the UK you need to be registered for National Insurance to be allowed to work legally) to be issued to migrants from the EU.[5] Despite warnings from both EU officials and UK business that such quotas would both be difficult to impose under EU law and hard to make workable, he allegedly is intending to keep pressing ahead with his plans as part of the UK's EU renegotiations (see below).

Mr Cameron is also attempting to renegotiate the UK's relationship with the EU with immigration as a key point. Yet in his widely trailed speech giving his definitive view on immigration on 28 November 2014,[6] he suggested rather more limited changes mainly affecting deportation of convicted criminals, shutting down sham colleges and more powers to detect sham marriages and restriction of migrants' welfare rights. One gets the strong impression that he blows hot and cold on the subject.

The pre-election Labour party leader Ed Miliband[7] has responded with plans, "to make sure migrants have to 'earn the right' to state benefits and face stiff English language tests before taking up jobs as well more policies to combat low wage exploitation".

Meanwhile, the Liberal Democrats, who have traditionally been more welcoming to immigration, used distinctly tougher language on immigration than had been traditionally the case, saying in their manifesto:

'Liberal Democrats believe Britain must be open for business and growth but closed to crooks and cheats. Britain needs more students and more visitors to come to help our economy grow. We will encourage people to visit Britain, learn in Britain and contribute to Britain. We will say yes to doctors, experts, entrepreneurs and investors. But we will say no to crooks, traffickers and those who would damage our country.

'By bringing back proper border checks - so we know who's coming in and leaving the UK - we will identify and deport people who over-stay their visa. We will create visible security and firm control, with real processes to count everyone in and count everyone out. No more guesswork on numbers: real evidence to catch out over-stayers. We'll ensure people can speak English and are willing to work. We'll ensure that migrants, including from the EU, come to work or study, not to claim benefits. And when it's time for them to leave, we will make sure they return home'[8.]

The fast-rising UK Independence Party Party focussed much of its attention during the election on immigration as well as its long running policy for the UK to leave the

EU and negotiate a free trade area with the remaining EU member states.

UKIP produced a comprehensive policy on immigration for the 2015 UK election. The key element was limiting net migration to 50,000 a year under an Australian-style points based system. This was to be backed up by a range of supporting measures as can be seen from the following extracts from the UKIP immigration manifesto :[9]

'UKIP will amongst other things:

• Create a Migration Control Commission – with a remit to bring down net immigration, and establish a visa system based on an Australian points based system while assuring the right number of highly skilled workers from across the globe are able to enter

• Increase Border Agency Staff by 2,500. Staff to be allocated in new division outside of current management structure

• Establish an ethical Visa system for work and study to be based on the principle of equal application to all people

• Abolish rules discriminating EU citizens from non-EU citizens

• No amnesty on illegal immigration

ON SECURITY AND ENTRY POINT SYSTEMS

• We would establish one passport queue for British Citizens and a second passport queue for Rest of the World;

• We would increase Border Agency Staff by 2,500, with staff to be allocated in a new division outside of current management structure

• We would implement new technology that is able to ensure all passport and visa holders are counted in and out and identifies over-stayers and we'd commence negotiations with other countries with biometric and improved technology for access to faster passport access to the UK.

ON EMPLOYMENT AND VISAS

- UKIP plans to establish an ethical visa system for work and study to be based on the principle of equal application to all people.

- UKIP plans to abolish rules discriminating against non-EU citizens in favour of EU citizens.

- UKIP plans to establish a visa system based on the Australian points based system.

- And UKIP plans a 'Highly Skilled Workers' only work visa, providing the opportunity for permanent leave to remain.

ON VISAS

- We'd have five tiers for visas

- These include the Highly Skilled Work Visa, the Temporary Unskilled Workers Visa, Visitors Passes, the Student visa, and the Family Reunion visa.

ON ASYLUM

- We'd maintain principles of UN Convention on Refugees for Asylum and have immediate review of the asylum process which aims to speed up rights to Leave To Remain and discover logjam on those declined asylum statuses.

ON ILLEGAL IMMIGRATION

- UKIP will offer no amnesty on illegal immigration.

- UKIP will increase the police enforcement team by 500 extra front line staff.

- UKIP will improve technology at police stations to ensure better identification of illegal immigrants between enforcement and arrest.

• And we'd review the current holding and accommodation for illegal immigrants.

ON CITIZENSHIP

• Save for current applications and Highly Skilled Workers, approved asylum and family reunion, there will be no permanent leave to remain permitted outside of the new policy rules

• Existing EU citizens will be offered the opportunity to seek permanent right to reside and citizenship in the UK when Britain leaves the European Union

• Criminals will lose the right of citizenship

• And we would remove the passports of those who choose to fight alongside terrorist organisations.

ON BENEFITS & HEALTH

•   UKIP says that all Highly Skilled Persons will be required to have Health Insurance for 5 years.

• UKIP says that all immigrants under the new, Australian-style points system will have to make tax and national insurance contributions for five consecutive years before claiming UK benefits.

• And UKIP wants our government to have a fund developed to cover those migrants who fall outside of the above or in NHS.'

Economics

Although politicians vied with each other to produce increasingly tough statements and policies – as exemplified by the details of the UKIP policy set out above, few economists or serious businessmen would support any noticeable reduction in immigration. Far from being a cost to society, economists generally see immigration as an asset.

John Cridland, former leader of the business representative organisation the Confederation of British Industry (the CBI),[10] has been quoted reacting to politicians proposals to limit immigration by saying that any such changes would have a "material impact" on UK competitiveness.[11]

The official report by the UK Department for Business, Innovation and Skills (BIS)[12] concludes, "there is relatively little evidence that migration has caused statistically significant displacement of UK natives from the labour market in periods when the economy has been strong. However, in line with some recent studies, there is evidence for some labour market displacement in recent years when the economy was in recession" and that "To date there has been little evidence in the literature of a statistically significant impact from EU migration on native employment outcomes, although significant EU migration is still a relatively recent phenomenon and this does not imply that impacts do not occur in some circumstances".[13]

A study published in *The Economic Journal*[14] concluded that between 1995 and 2011, immigrants from the European Economic Area (EEA) contributed £8.8bn more than they received in benefits. That compared with a drain on the nation's finances of £604.5bn by native Britons[15]. More recent migrants contributed even more, the study found: "Our analysis suggests that rather than being a drain on the UK's fiscal system, immigrants arriving since the early 2000s have made substantial net contributions to [the UK's] public finances, a reality that contrasts starkly with the view often maintained in public debate," the report concludes. "Between 2001 and 2011, recent EEA immigrants contributed to the fiscal system 34 % more than they took out, with a net fiscal contribution of about £22.1bn," the study said. Migrants from non-EEA countries made a more modest but still positive contribution, putting in roughly 2% more than they took out.

But over the same period, native Britons paid much less into the system than they took out in benefits, contributing only 89% of what they cost. Even when the data are adjusted to take account of differences with the native population in gender mix, age and educational attainment, recent immigrants are still 21% less likely than natives to be receiving benefits.

A separate study by the UK-based National Institute for Economic and Social Research[16] looked at the effects of immigration on worker productivity. This study concluded that those industries with higher shares of migrant workers had higher labour productivity. The correlation was particularly noticeable in the manufacturing and real estate sectors.

"Our analyses show a positive and significant association between immigration and labour productivity growth in the time period analysed," the study found, noting that each 1.0% rise in the immigrant share of employment is associated with an increase in labour productivity of 0.06 to 0.07%[17]. Yet these studies do not seem to capture the key elements of the impact of immigration on the UK economy and particularly on the London economy.

A dynamic framework for analysing the economic impact of immigration

To understand the full impact of immigration on economic growth, one needs to treat it as a dynamic process that not only has a direct effect, but a substantial enabling effect to boost other economic processes that are taking place at the same time.

My analysis of the economic impact of immigration suggests a threefold effect:

- First: migration alleviates skill bottlenecks and hence removes barriers to faster growth.
- Second: migration boosts diversity and also boosts creativity which enhances productivity with both direct and indirect effects on economic growth.
- Third: migration makes business more profitable, enhancing investment and hence growth. Even if the initial impact is to place downward pressure on wages through the enhanced supply of labour relative to demand, the secondary effect is to boost wages through faster growth.

Skill bottlenecks

The CBI together with KPMG used to run a regular survey of London's business sector. The September 2014 survey[18] quoted 45% of respondents as claiming that their rate of growth is likely to be curtailed by shortages of skills. And two-thirds of the respondents (66%) reported difficulty in recruiting skilled staff. The areas of greatest shortage included IT and technology specialists, creative specialists, while those with finance and engineering skills were also in high demand. The survey also indicated lacks of basic skills such as core literacy and numeracy, communication skills and business awareness.

Unless the respondents to the survey are completely wrong in their assessment, these results indicate the extent to which – even after substantial flows of immigrants into London, skill shortages remain. Moreover, the most pressing of these shortages are in the key areas for the Flat White Economy – IT and technology skills and creative skills. This is compelling evidence that migration – to the extent to which it alleviates bottlenecks – is likely to have a disproportionately positive effect on economic growth. Since 2014 the CBI have changed the survey (and taken on a new sponsor in the property consultancy CBRE) and more recent comparable data does not appear to be available.[19]

Given that even after migration these skill bottlenecks still existed in 2014, it is frightening to think how considerable they might prove if the safety valve of immigration were to be screwed shut. Note that the specific question to which the businessmen were responding relates to what is currently slowing down their growth and that as high a proportion as 45% claimed that even in 2014 their growth was constrained by lack of skills. Having run such surveys myself, my experience is that the types of businessmen who answer them give an impressionistic view which is sufficiently accurate[19] to be taken seriously.

Migration, diversity and creativity

We examined in Chapter 2 the link between migration, diversity and creativity. This is very evident in the arts but we showed how similar forces are important in sectors as diverse as marketing and advertising and in technology. Most employers find benefits from diverse labour forces with employees with different backgrounds stimulating the thinking of each other.

The research by NIESR quoted earlier in this chapter shows the correlation between migration and productivity and it seems likely that the boost to creativity from migration is a significant component to this. We also showed in Chapter 2 that migrants to London were one and a half times more likely to possess a degree or higher qualifications than those migrating to elsewhere in the UK. We also quoted a major director from the retail giant Walmart pointing out that the diversity resulting from migration encourages "more ideas and perspectives into driving the best business solutions".

And, there is academic research that appears to support this such as that by Professor Ronald Burt at Chicago University. Admittedly old age encourages one to be sceptical about 'proving' anything through academic research. But in this case, my own experience as a successful entrepreneur supports the conclusion that diversity helps create better business solutions, which is one of the reasons I try to maintain a highly diverse office. Actually in London one doesn't have to try too hard because the range of people making themselves available for any job is bound to be diverse now because of the diverse population in the capital. This very diversity is part of the London advantage which has driven the growth of the Flat White Economy.

Profits and growth

Rather more controversial than the likely impact of migration on skills and on creativity is the potential impact on profits and growth. Here we need to take a long discursion into economic theory before coming back to the issue of immigration itself. I think it worthwhile to look at this argument, because traditionally in the UK the Left have been in favour of immigration and the Right against. What emerges here is a relatively novel argument – a right-wing case for immigration based on the evidence that immigration boosts profits, which in turn boosts economic growth.

I covered this issue when I was the Gresham Professor of Commerce in my lecture 'Was Karl Marx Always Wrong?'. This lecture used Marx's Labour Theory of Value to postulate that the growth in so-called 'exploitation', as the world's effective labour supply had increased, was in fact improving the outlook for employees compared with the hypothetical alternative.

The lecture argued that workers who were initially 'exploited' in the sense of being paid relatively low wages – but whose real pay increased faster as a result of high profits leading to a high rate of economic development – would in general be better off than workers (who in theory were less 'exploited' initially) for whom real pay increased more slowly as a result of lower profits and less economic development.

It is a paradox that the 'exploited' worker often does better than one who is not. But obviously not in all circumstances...

A worked example

In my lecture I showed with a simple worked example how wages would move in different theoretical situations.

The assumptions are these:

- GDP is divided into wages and profits.
- Profits are wholly invested.
- There is a constant capital output ratio.
- So the higher the profits the greater the addition to GDP.

Obviously these are highly stylised assumptions but in fact relaxing them to bring them closer to real world experience does surprisingly little damage to the conclusions.

The results are shown in Figure 7.2 for different initial starting positions where wages as a share of GDP are allowed to vary from 20% to 90%. What this analysis shows is that if wages are too low no matter how fast they grow, they remain low. And if wages are too high, the workers lose out because profits are too low to finance investment and hence economic growth dries up. The sweet spot for wages is between 30% and 60% of GDP. The sweetest spot on the graph over a ten-year horizon is for wages to be slightly below that of 50% of GDP.

Actually this was all anticipated (without the mathematics) by Marx. He postulated that capitalism would

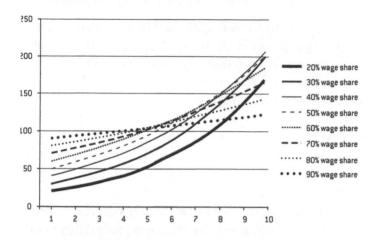

Figure 7.2: Evolution of Wages Under Different Levels of 'Exploitation': labour does best when its share is between 30% and 60%.

"die of its own contradictions." What he meant by this was that workers' resistance to capitalism would force wages up to such a high share of GDP and squeeze profits to such a low level that investment would dry up, growth would slow and possibly go negative since a certain level of base investment is necessary simply to replace the capital stock. He argued that this would cause an economic collapse that would in turn cause the proletarian revolution which would lead to communism.

And actually it nearly happened in the UK. The inflationary crisis of the late 1960s and early 1970s forced down the rate of return on capital employed to around 2% by 1975 and the profits share measured properly down to nearly 20% of GDP (the official numbers were higher than this but included imaginary inflationary gains).

I've always had a soft spot for Marx for two reasons. First, he put profits at the heart of his analysis of the capitalist system. Far too many conventional economists fail to understand the role of profits and assume they just happen. Is it too fanciful to detect the hand of Engels, the Manchester businessman, in Marx's serious appreciation of how profits are intrinsic to the capitalist system? Second, he realised that economics didn't just happen in a vacuum and integrated his social and political theories with his economics.

My Marxist supervisor for my Master's degree, Andrew Glyn, wrote a book about the decline in profitability in the UK in the 1960s and 1970s.[21] He argued that the collapse in profitability under pressure from the trade unions might possibly be the beginnings of the period when capitalism collapsed in the UK. I had just joined the CBI shortly after reading Glyn's analysis and interestingly many businessmen shared this Marxist analysis that capitalism might be about to collapse in the UK.[22]

Glyn was not far wrong – it is possible to argue that only the chance combination of Mrs Thatcher, the split in the Labour Party as the SDP was formed in the early 1980s, the Falklands War and General Galtieri, the distaste of the British public for trade union excesses, the emergence of the new technologies of the 1980s and the beginnings of globalisation combined to prove him wrong.[23] But it was a close run thing and without a remarkable combination of forces emerging at the same time he could easily have been right.

Although the 1980s recovery of capitalism looks more like a politically led bounceback, since the 1980s the recovery of capitalism has been due more to technology and globalisation.

What has been happening to profits?

The distinguished economist Nicholas Kaldor who was one of Harold Wilson's economic advisers (and someone whom I remember vaguely from my days as a postgraduate student, though he wasn't the sort of person to bother with someone who was then as insignificant as me) summarised the statistical properties of long-term economic growth in an influential 1957 paper.[24] He pointed out six "remarkable historical constancies revealed by recent empirical investigations". The most important of these is his assertion that the "shares of national income received by labour and capital are roughly constant over long periods of time".

His broad generalizations, which were initially derived from U.S. and U.K. data, but were later found to be true for many other countries as well, came to be known as "stylized facts".

In my Gresham lectures I have drawn attention to the occasional operation in economics of what the sophisticated call Heisenberg's uncertainty principle or what the less sophisticated call Murphy's Law or Sod's Law: that when you start observing a variable it starts behaving differently. And the moment you draw attention to a variable that has remained constant for a long period the variable – of course – starts behaving differently. And indeed the data on labour shares of income has started to behave differently since Kaldor made this observation[25.]

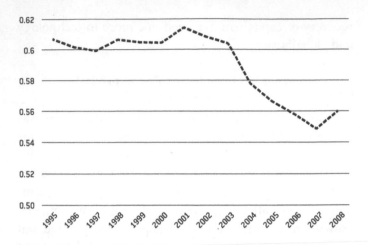

Figure 7.3: The fall in the workers' share of advanced economies' incomes.

Because the labour share of income can be measured in a variety of ways, I've looked at various different studies and taken the conclusions from each of them. What they all show is that there has been a sharp fall in the proportion of total income going to labour (and a corresponding increase in the return to capital).

The McKinsey Global Institute[26] claims a 7.1% fall in the share in the labour share of income since its peak in 1975 for "advanced economies". This is backed up by more recent data from the OECD shown in Figure 7.3 that shows a 4.7% drop from 1995 to 2008.[27]

Figure 7.4 is from the OECD, via *The Economist*'s edition of the 2nd November 2013, where this pattern was first highlighted. The OECD data shows the labour share of income edging down – also since the mid-1970s – in

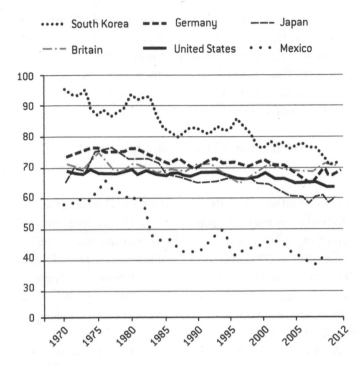

Figure 7.4: Labour costs as a share of GDP. Source: wwww.oecd.org[28]

South Korea, Germany, Japan, Britain, the US and Mexico.

Meanwhile a very comprehensive study from the National Bureau of Economic Research in the US[29] shows the data for the world's four largest economies – US, China (though only for a relatively short period), Japan and Germany. It again confirms the downward movement in the labour share.

It is also interesting that the three emerging economies for which we have data in this sample – China, Mexico and South Korea – have much lower labour shares of income than in the advanced economies (which might be a bit of evidence to support Marx's contention that capitalism in the long term would ultimately bid profits down to a level that is too low to permit economic growth). However, these emerging economies currently have much faster rates of economic growth.

What is the evidence that higher profits boost economic growth?

A crude statistical analysis for the OECD economies for which data is available shows a negative correlation coefficient of -0.31 between the labour income share average from 2000 to 2006 and the rate of economic growth from 2001 to 2008. What this says is that there is a statistically significant negative correlation between the labour income share and GDP growth. In other words, the higher the share of profits, the faster the rate of growth.

Although the capital theorists differ, two undoubted heavyweights in Karl Marx and John Maynard Keynes definitely believed that the higher profits and hence investment, the faster the rate of economic growth. Here for example, Keynes in probably his most readable book, *The Economic Consequences of the Peace*,[30] wrote about the pre-1914 world which had experienced a bout of globalisation very similar to that of recent years:

'Europe was so organized socially and economically as to secure the maximum accumulation of capital. While there was some continuous improvement in the daily conditions of life of the mass of the population, Society was so framed as to throw a great part of the increased income into the control of the class least likely to consume it. The new rich of the nineteenth century were not brought up to large expenditures, and preferred the power which investment gave them to the pleasures of immediate consumption. In fact, it was precisely the inequality of the distribution of wealth which made possible those vast accumulations of fixed wealth and of capital improvements which distinguished that age from all others [...] The immense accumulations of fixed capital which, to the great benefit of mankind, were built up during the half century before the war, could never have come about in a Society where wealth was divided equitably. The railways of the world, which that age built as a monument to posterity, were, not less than the Pyramids of Egypt, the work of labour which was not free to consume in immediate enjoyment the full equivalent of its efforts. [...] Thus this remarkable system depended for its growth on a double bluff or deception. On the one hand the labouring classes accepted from ignorance or powerlessness, or were compelled, persuaded, or cajoled by custom, convention, authority, and the well-established order of Society into accepting, a situation in which they could call their own very little of the cake that they and Nature and the capitalists were co-operating to produce. '

Marx's explanation is less clearly put, but he is in effect making the same point:

'The capitalist shares with the miser the passion for wealth as wealth. But that which in the miser is a mere idiosyncrasy, is, in the capitalist the effect of the social mechanism of which he is but one of the wheels ... [T]he development of capitalist production makes it constantly necessary to keep increasing the amount of the capital laid out in a given industrial undertaking, and competition makes the . . .laws of capitalist production to be felt by each individual capitalist as external coercive laws. It compels him to keep constantly extending his capital, in order to preserve it, but extend it he cannot, except by means of progressive accumulation[31].'

I think it impossible to prove absolutely that higher profits shares boost economic growth (and indeed when the share of profits becomes extreme on the high side, one suspects that growth is depressed through lack of demand), but given that two of the greatest economic theorists (neither of whom are generally assessed as particularly right wing) believed it to be the case and that the empirical evidence is supportive, one would have to find a remarkable amount of contradictory evidence, that to the best of my knowledge is unavailable, to mount a convincing case in the opposite direction.

What if the capitalists blow the profits on high living?

Of course the capitalists might well take high profits out

of their businesses as dividends and then spend rather than reinvest them. There are enough examples in history of this for the possibility not just to be theoretical. Yet the correlation between low labour income shares (implying high profitability) and economic growth is a compelling indication that it is more likely than not that profits get reinvested to generate growth. This is because when business is highly profitable, it not only provides funds for investment but also makes the returns on investment attractive. Equally, high rates of profitability make external financing easier.

The UK profits share is still too low

But the declines in labour income shares in some of the Western economies like the UK still leave them well above the sweet spot. In the UK for example, the OECD figure for the labour share, although lower than the disastrously high 80% share in the mid 1970s, was 65.8% in 2010 according to OECD figures. It is because of this that a boost to the labour supply such as that caused by migration that depresses the share of wages initally has potentially beneficial dynamic effects.

In recent years the UK economy has benefitted from two labour market shocks that have boosted the supply of labour. The first and best known has been from immigration. The second, which is much less publicised has resulted from the underfunding of pensions which has forced people out of early retirement and back into work. In the 21st century a third of the increase in people in work in the UK has resulted from increased numbers of pensioners having to work to augment their pensions. In July to September 2015 there were 1,168,000 people over the age of 65 in the UK in work. In 2001 the equivalent number was only 437,000. These workers are probably not in general available for the Flat White Economy but they are an unappreciated addition to the UK labour force as a whole.

It is conceivable that Gordon Brown was a Marxist who understood what he was doing – boosting the labour supply by changing immigration rules and imposing heavy taxes on pensions that forced people to work in their old age. And certainly I know from personal experience that some of his advisers are, or at least were when younger, Marxists.

But – though it would be fun to believe otherwise – I suspect that the cock-up theory is a better explanation than the conspiracy theory. I doubt if Brown thought that profits were too low and that he had to boost them by force-feeding the economy with labour. I suspect he contributed to the wrecking of the pensions system by accident in an attempt to raise money and was pleasantly surprised to discover the boost to the labour supply that resulted. The conspiracy theory works slightly better in the case of immigration because the Labour Party tends to support immigration because it thinks that immigrants vote for them, but even then, I doubt if the government really intended the boost to immigration that has taken place. In any case, the pace of immigration has if anything accelerated under the Coalition who would be unlikely to be conspiring to cause it for political reasons.

But what is clear is that both – and in the context of the Flat White Economy particularly immigration – has helped keep labour supplies up and has made this economy more profitable for entrepreneurs. If we accept my argument that profits in the UK have traditionally been too low, then the boost to the labour supply from immigration is a major component in boosting the economy through depressing wages and boosting profits.

Immigration profitability and growth

Because of the link between profits and growth, which is supported by the data and the analysis, to the extent that immigration increases profitability, by limiting wages in a given growth environment, it both acts as a signal for investment and provides the funds to support such investment. This is especially important for small businesses such as those in the Flat White Economy.

Small businesses drive a substantial proportion of economic growth (a recent report by Cebr for Octopus shows that 1% of SMEs generate 68% of the UK's economic growth)[32]. Small businesses traditionally have difficulty in raising funds externally. This increases the vital importance of the funds generated internally. Also, to the extent to which it is possible for them to raise funds externally, internally generated funds greatly increase the potential for this[33].

It may be unfashionable to say so, but to the extent to which immigration keeps wages down initially, *ceteris paribus*, it permits higher growth and higher wages eventually as a result of the positive dynamic effects through boosting profitability, investment and hence growth.

Negative effects of immigration

Of course not all impacts from immigration are positive. Immigration can have an adverse impact on social cohesion. It can place strains on public services and infrastructure. In a crowded city like London, it is likely to create a housing shortage. But these are manageable. Even the static analysis of the impact of immigration on public finances, which takes no account of the boost to economic growth from higher profitability, or from the other dynamic effects, shows a massive public finances surplus which could easily absorb the costs of improved infrastructure and public services to cope with the requirements of the migrants which are typically relatively low because of their age distribution.

Housing is more of a problem and London has a chronic housing shortage. But the evidence from *The Economic Journal* article seems to be that migrants use much less housing stock per capita than other members of the population and typically place the strain on the private rather than the public sector.

Conclusions

So immigration has a very high-powered impact on the economy. It supplies skills and hence alleviates skill shortages. It boosts diversity and creativity – an office with employees from a diverse range of cultures tends to think 'outside the box' and moves beyond pre-packaged solutions. And it boosts profitability and growth. Paradoxically, even if immigration initially depresses wages, its dynamic effects act in the opposite direction. This is why academic studies show that immigration does not appear in aggregate either to depress wages or employment for the domestic population.

For the Flat White Economy, the advantages of migration are even more important. This is an economy based on technological skills and creativity, which are the skills in shortest supply and those boosted most from migration. To sum up, it is worth pointing out that it is not only London which believes it benefits disproportionately from migration. In recent years civic leaders from both New York and Toronto have made similar assessments.

It would be surprising if the leaders of all three cities were wrong.

CHAPTER 8

The UK Economy: the Flat White Effect

It will be no surprise to readers that the UK economy is concentrated on London. What might be a surprise is the extent to which this is so and the even greater extent to which the UK's economic dynamism is based in London. So the Flat White Economy has been the most important single motor driving the UK economy in the 2010s. George Osborne's economic recovery has been founded on it and the shift in London's employment pattern from highly paid and highly taxed City jobs to less highly paid and much less highly taxed jobs in the Flat White Economy is one of the reasons why the economic recovery has not reduced the government's deficit by as much as conventional forecasters had expected.[1]

This chapter looks in detail at the impact of the Flat White Economy on the UK as a whole and how the UK depends upon it as a driving force. It provides new figures which show that the contribution of the sector to the national economy is understated by existing statistics and adjusts the UK growth figures to correct for the understatement. But most importantly it analyses the key forces behind economic growth in a modern economy and shows how the Flat White Economy has become the critical element in the UK's recent transformation as an economy. Again, it shows new figures that indicate that official government indicators seriously understate the scale of innovative activity taking place in the UK and so fail to appreciate the extent to which the digital economy is driving the UK forward.

Input Output statistics

There is a very detailed source of economic information, of interest normally only to the purists, called 'input output statistics'[2]. These split the economy into over one hundred sectors and show how each of these sectors trade with each other as well as selling goods and services for final use to the consumer, investor, to the government and for overseas trade.

Most government statistics for the economy treat the economy as if it were vertically integrated as one company, a sort of 'UK plc'. For many purposes this is a useful simplification, but not always. The input output statistics vertically 'disintegrate' the economy, showing the messy detail of business-to-business sales. This is becoming especially important in modern business models as sub-contracting is a key business asset and having high quality suppliers is critical for companies.

A modern economy has a strong business-to-business sector and the UK has an especially high ratio of business -to-business sales compared with other countries; the only studies for this date back to the 1990s but they support the case. This aspect of modern business is part of the modernisation of the UK economy that started in the 1980s. The Flat White Economy is a continuation of this trend. It is particularly important as the whole economy becomes ever more digital. Digital expertise is mostly supplied by specialists; it is at the core of the Flat White Economy.

The first thing to note is the sheer scale of the Digital Sector. Even in 2013, the latest year for which the latest Input Output data is available, the computer programming and consultancy sector was the eighth largest private sector industry in the UK after construction, retail, financial services, wholesale, property, hotels and catering and insurance. There are also three parts of the public sector which are larger: the Civil Service and local government; education and the health service and one additional sector which gets included for statistical reasons – the imputed rental value of the buildings that owner occupiers are deemed to be renting to themselves.

I doubt if many people realise that computer services and programming is twice the size of the UK oil and gas sector and roughly twice the size of the UK car industry.

If you add together all the sectors of the Flat White Economy including cultural, film, advertising and marketing and the IT sector (even if you exclude telecoms) the sector's value of UK supply at basic prices in 2013 amounted to £135.0 billion, which would have made it the second largest UK sector in the private sector after construction, just slightly larger than both retail and financial services (excluding insurance). It is actually already nearly the same size as the whole of financial services (in other words not only the City but also the high street banksand

Table 8.1: Value at basic prices (UK, £millions) in 2013 – all UK sectors with UK output more than £100 billion

Sector	Value of UK supply at basic prices
Construction	221,893
Owner-Occupiers' Housing Services	148,894
Human health services	144,819
Public administration and defence services; compulsory social security services	144,103
Retail trade services, except of motor vehicles and motorcycles	128,537
Financial services, except insurance and pension funding	125,923
Education services	122,709
Wholesale trade services, except of motor vehicles and motorcycles	115,873
Real estate services, excluding on a fee or contract basis and imputed rent	84,167
Food and beverage serving services	76,990
Insurance and reinsurance, except compulsory social security & Pension funding	71,963
Computer programming, consultancy and related services	69,622
Residential Care & Social Work Activities	69,044
Electricity, transmission and distribution	61,160

the people who sell life insurance, (though not acutal insurance) and pensions or the whole of wholesale. The contribution of the Flat White Economy defined in this way in 2013 was 8.7% of GDP at basic prices.

Our forecast is that by 2025 this contribution will rise to 15.8% of GDP, which will make it the UK's single largest business sector. This is an extraordinary result for part of what in 1964 the soon-to-be Labour Prime Minister Harold Wilson[3] called the "candyfloss economy". The true scale of this sector is often not fully understood. But there is an accounting quirk which means that its full contribution to the UK economy is even then still actually understated.

Accounting for IT spending

When you buy IT, you have some choice about how to account for it. The capital equipment normally has to be treated as investment, but things like software can be either capitalised (put down as investment rather than as a cost of sales) or treated as a running expense. If you capitalise it you get an investment allowance or a depreciation allowance. For small investments you can get a 100% allowance (the Annual Investment Allowance), but for larger investments, if you capitalise the expenditure, you only get a depreciation allowance. Suppose you depreciate over five years. In effect you get a tax allowance against profit of 20% each year. But if you expense the software you can get in effect a 100% investment allowance by reducing your taxable profits. So this tax treatment encourages you to expense software.

The treatment encouraged by accountants also acts to discourage you from capitalising software. There is some sense in this since one factor which weighs heavily with accountants is what the assets of the business would be worth if they had to be sold up. Since most software is worthless if the business shuts down, they are unenthusiastic about capitalising such investment.

The section below from a well-known accounting website sets out the relevant rules:

The relevant accounting is:

- Stage 1: Preliminary. All costs incurred during the preliminary stage of a development project should be charged to expense as incurred. This stage is considered to include making decisions about the allocation of resources, determining performance requirements, conducting supplier demonstrations, evaluating technology, and supplier selection.
- Stage 2: Application development. Capitalise the costs incurred to develop internal-use software, which may include coding, hardware installation, and testing. Any costs related to data conversion, user training, administration, and overhead should be charged to expense as incurred. Only the following costs can be capitalised:

    – Materials and services consumed in the development effort, such as third-party development fees, software purchase costs, and travel costs related to development work.

    – The payroll costs of those employees directly associated with software development.

    – The capitalisation of interest costs incurred to fund the project.
- Stage 3. Post-implementation. Charge all post-implementation costs to expense as incurred. Samples of these costs are training and maintenance costs.

Any allowable capitalisation of costs should begin after the preliminary stage has been completed, management commits to funding the project, it is probable that the project will be completed, and the software will be used for its intended *function*.

You can see from this that the accounting profession (not without good reason) is pretty unenthusiastic about the investment being capitalised. The economic accounting perspective is not quite the same. For an item to be categorised as investment it only has to yield a return after the accounting period. The issues of going concern and breakup value are not really relevant in economic accounting as opposed to financial accounting.

In the classic work on economic accounting by the OECD,[4] investment is defined as:

> *In the national accounts, investment, i.e. the purchase of machinery (including software) and buildings (offices, infrastructure, dwellings) and the constitution of stocks (inventories) is known as gross capital formation (GCF).*

This implies that most if not all of software expenditure should, in economic terms, be treated as investment. This is confirmed by the Bank of England guidance[5] on how software should be categorised as investment:

> *Within the 'Plant and machinery' category, 'Computer hardware' includes expenditure on microcomputers, printers, terminals and optical and magnetic readers, including operating systems and software bundled*

with micro-computers purchased. 'Computer software' encompasses software licence payments and all capitalised items of computer software consultancy and supply, including the purchase or development of large databases. Expenditure on 'In-house software development' includes the costs, including staff costs, of developing software with a useful life of at least one year.

Yet when you look at how software is actually treated in the National Accounts, the figures from the 2013 input output tables show that while £35.0 billion was allocated to final demand (almost entirely investment), £46.1 billion were allocated to intermediate demand. Since virtually any software investment must have a useful life of at least one year, it is hard to see how approximately 60% of software expenditure should be expensed or allocated to intermediate demand from an economics perspective, even if such a practice would be justifiable from an accounting perspective. Indeed it is hard to see how any software expenditure other than routine maintenance should be allocated to so-called 'intermediate demand'.

Now the importance of this may not be obvious to a non-economic technician, but when output is allocated to intermediate demand it is deducted from the purchasing sectors' contribution to GDP. As a result, allocated to intermediate demand it has no net impact on GDP whereas when it is treated as investment it is treated as a positive contribution to GDP from its own sector and not deducted from any other sector.

Since we have shown pretty categorically that it should not be allocated to intermediate demand, we are looking at a basic understatement of UK GDP of about 3.0%. This excludes any additional understatement from such treatment of other parts of the Flat White Economy such as related investment in marketing and advertising.

Table 8.2: Understatement of GDP 1997–2012

Year	Expensed Software (£million at basic prices)	% of GDP at basic prices	Understatement of GDP growth
1997	8,773	1.11%	
1998	12,786	1.55%	0.44%
1999	17,003	1.98%	0.43%
2000	20,082	2.20%	0.22%
2001	21,588	2.28%	0.07%
2002	26,570	2.66%	0.38%
2003	30,711	2.88%	0.22%
2004	33,836	3.01%	0.13%
2005	33,918	2.85%	-0.16%
2006	38,619	3.07%	0.21%
2007	40,286	3.03%	-0.03%
2008	42,042	3.07%	0.04%
2009	42,375	3.15%	0.08%
2010	41,694	2.98%	-0.17%
2011	45477	3.15%	0.18%
2012	44613	3.02%	-0.13%
2013	46164	2.98%	-0.05%

Table 8.2 shows how the failure accurately to measure software resulted in an understatement of GDP of roughly 3% for the years from 2004 to 2013, although the degree

of understatement remained roughly constant over the period after 2003. In the period from 1997 to 2004 the degree of understatement rose from 1.1% to 3.0%, leading not only to an understatement of GDP, but also to an understatement of GDP growth of slightly under 0.3% per annum over the period. Most other countries likewise account software expenditure in the same way – the difference being that in the UK the FWE sector is larger and the mismeasurement more significant.

IT investment as an enabling technology

In the 1980s, economic studies seemed not to be able to find much evidence of information technology making much difference to economic growth. This was known at the time as the 'productivity paradox.'[6] Nobel laureate Robert Solow famously quipped "You can see the computer age everywhere except in the productivity statistics".[7] But micro studies since have provided compelling evidence that IT is not only a technology that enhances growth but also one that enables further productivity gains. A Google search lists over 64,000 references for 'information technology as an enabler'.[8]

The modern thinking about IT investment is that it encourages improved and innovative products, services and methods of production. It improves organisation and quality. And so investment in IT means that growth is boosted in the UK by a range of different factors. Indeed an official UK government report by the Department of Business, Innovation and Skills (BIS)[9] indicates this. The analysis quoted from the Gröningen Growth and

Development Centre shows that between 1997 and 2007 more of UK growth was driven by ICT investment than in the US, France or Germany (see Figure 8.1).[10] Although, to be fair, you might expect the CEO of O2 and Telefónica in the UK, Ronan Dunne, to say this, Dunne has made it very clear that he expects the key growth driver overall to be information and communications technologies (ICT) and investment. In a speech to Microsoft's Future Decoded conference,[11] he explained how the next wave of digital will be controlled by mobile and cloud technologies converging, "but it's our children who will need to drive this change forward".

He further argued that the UK and European economy's future growth would be driven by a digital revolution. "We may in fact look back in time when we went from BC

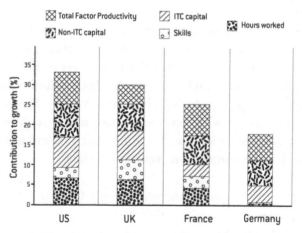

Figure 8.1: This chart shows that in the UK ICT Capital makes a proportionately larger contribution to economic growth than in any of the other countries.

'before connectivity' to AD 'after digital' and that is being driven by innovation and social media." He described connectivity as the oxygen of modern life for consumers, whilst stressing telecommunications are a major part of the country's infrastructure in the economy. "More people on the planet have access to a mobile phone than clean drinking water or a toothbrush," he explained, based on information taken from an Ofcom report.[12]

Meanwhile, the economists in the UK government's economic service have carried out some research into the importance of innovation (largely from ICT) in driving the UK's economic growth.[13] The report shows a diagram of how innovation propels growth on its first page which is too complicated to be easily reproduced[14] but the elaborate detail of it captures what is a realistic point: the routes through which innovation affects the economy are multifarious.

It starts off – remember that it is written by people who work for the government and whose essential political bias is to find ways of justifying additional ways of spending money, and who by and large earn too little to understand the full damage done by high levels of taxation – by showing how there is what economists call market failure in the area of innovation (market failure is a standard justification for government intervention). It states:

Innovative activity generates knowledge which has the characteristic of a public good. It "spills over" to benefit individuals and firms who did not make the initial investment. Because these benefits are not taken into account by the investor, society as a whole under invests in innovation activity. Investments in innovation are highly uncertain and may also involve long developmental timeframes. Investors often cannot quantify the likely returns to an investment, which may lead to inefficiency and difficulty in distinguishing good investments from the bad. Under-investment in innovation may therefore occur.

In spite of the possible bias from the authors, they are definitely correct in their first point and quite possibly correct in their second. But of course the Flat White Economy has emerged, generating most of the UK's innovation without any direct public intervention, though it has been the (probably unintended) beneficiary of planning and transport policy. Nevertheless, the case for public sector support for innovation can be turned on its head because it is the same case that essentially says that private sector led innovation in the Flat White Economy has had a beneficial effect on the whole UK economy that extends well beyond the direct returns to those involved. The modern economic literature concentrates not only on the importance of innovation itself, but also on its diffusion throughout the economy and on the economy's absorptive capacity. The most widely quoted book on the subject is by Everett Rogers, a professor of communication studies, who popularised the theory in his book *Diffusion of Innovations*. The book was first published in 1962, and is now in its fifth edition (2003).[15]

The Flat White Economy ticks all the boxes – much of it is a self-generated addition to business knowledge that reinforces commercial potential. It diffuses very rapidly for a number of reasons – there is a real community of techies and knowledge spreads very quickly between them. And there is huge absorptive capacity in the UK because of the underpinning of the extraordinarily high percentage of UK businesses that have moved their activities online and because of the UK's predominance in digital marketing.

Defining R&D – the enduring legacy of Frascati

The conventional wisdom is that the UK is good at innovation but pretty appalling at its commercial exploitation. However, this view is based on the so-called 'Frascati definition of R&D'. This is the conventional formulation of research and development (which is the main component of innovation) based on a definition agreed at a presumably rather enjoyable scientific conference in the Villa Falconiere in the Italian wine making town of Frascati, near Rome. I don't want to be too rude about the scientists (though occasionally I cannot resist the suggestion that the definition emerged after the consumption of a few too many glasses of the local vintage – hence the name). The problem may owe even more to the fact that those involved met in 1963 at a time when the internet, let alone the digital economy, had not even been invented. Their definition of R&D, and hence innovation, is constructed in such a way that most software development is consequently excluded.[16] The reason for this is that the scientists involved in the process (and the distinguished economist Christopher Freeman, emeritus Professor of the Economics of Science at Sussex University) were focussing on research carried out by the type of scientists operating in dedicated research establishments and wearing laboratory coats. In the days when this definition was invented, no one had ever envisaged that economically valuable research would be carried out by young(ish) kids playing at their desks with their computers in between other activities. But in the new digital economy this is how R&D mainly happens, and with the information economy being one of the largest parts of the world's economy and certainly its fastest growing, this probably accounts for the bulk of the innovative activity that takes place anywhere in the world, and especially in the UK. The result is that the desk research conducted by government economists using the Frascati definition of research misses the key points – that the UK is not only one of the leading nations in the world for innovation but also for its dissemination. Even if the scientists had not drunk too many bottles of Frascati (wine), they could not have produced a much more misleading indicator even if they had.

The official government estimate for UK R&D spend (in 2011) was £27.4 billion from the government's official 'Innovation Report 2014'.[17] Yet this is dwarfed by the £72.8 billion of computer programming etc (estimated for 2013), let alone the £135.0 billion for the whole Flat White Economy. Certainly the bulk of the first item of R&D *is* likely to comprise innovative activity. Yet there is likely to be an awful lot more to take into account than that. Even if we just allow a very conservative 50% of the activity of only the computer programming activity to be counted as innovation, it more than doubles the UK's total for estimated R&D spending to a huge £62.5 billion[18] in 2011.

Yet the government's Innovation Report suggests that

innovation accounts for "up to 70% of economic growth in the long run". The implication therefore is that the Flat White Economy is the key factor driving the UK economy forward, accounting for about 40% of all UK economic growth.

Multiplier effects

All these estimates do not of course directly take into account the multiplier effects of the Flat White Economy.

I have not yet been able to estimate the multiplier effect of the Flat White Economy as such. But since there is considerable overlap with the 'Creative Economy' that has been analysed by Cebr for Falmouth University,[19] the analysis contained in this report should give a pretty reasonable rough estimate. The process is shown in the diagram for Figure 8.2.

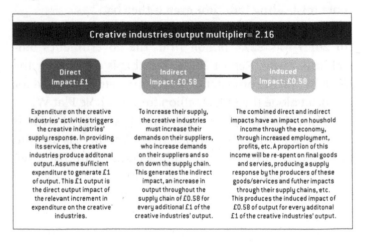

Creative industries output multiplier= 2.16

Direct Impact: £1	Indirect Impact: £0.58	Induced Impact: £0.58
Expenditure on the creative industries' activities triggers the creative industries' supply response. In providing its services, the creative industries produce additonal output. Assume sufficient expenditure to generate £1 of output. This £1 output is the direct output impact of the relevant increment in expenditure on the creative industries.	To increase their supply, the creative industries must increase their demands on their suppliers, who increase demands on their suppliers and so on down the supply chain. This generates the indirect impact, an increase in output throughout the supply chain of £0.58 for every additional £1 of the creative industries' output.	The combined direct and indirect impacts have an impact on houshold income through the economy, through increased employment, profits, etc. A proportion of this income will be re-spent on final goods and servies, producing a supply response by the producers of these goods/services and futher impacts through their supply chains, etc. This produces the induced impact of £0.58 of output for every additonal £1 of the creative industries' output.

Figure 8.2: The multiplier effect of the creative industries

To quote from the Cebr report:

- Through the input output modelling process, we also quantified the output multiplier for the creative industries. This is detailed in Figure 8.2 opposite.
- The creative industries' Type II multiplier is 2.16. This means that for every £1 of additional output in the creative industries, the economy-wide increase in output through direct, indirect and induced multiplier impacts is 2.16.
- Based on this multiplier effect, the creative industries' output contribution to the UK economy of £141 billion in 2012 is consistent with an aggregate output contribution, including indirect and induced multiplier impacts, of £304 billion.

The implication of this modelling is that the £135.0 billion value added of the Flat White Economy has had beneficial indirect effects and induced effects that generated £2491billion of GDP in 2013 – 18.8% of GDP. And if the multiplier were to remain the same, by 2025 around a third of the whole UK economy will be related to the Flat White Economy (a forecast 15.8% of GDP value added at basic prices with a 2.16 multiplier).

Conclusion

What this chapter shows is that the Flat White Economy is many times more important than is conventionally assumed. Its overall contribution to the economy is understated to the extent of over £40 billion. Its contribution to economic growth is also understated because of Frascati's international scientists' rather exclusive definition of R&D activity which in the 21st century massively understates what is actually taking place in the real world. And – as pointed out above – even on a conservative estimate of the impact on innovation of the Flat White Economy – it accounts for not far short of half of UK economic growth. And if the multiplier effects are taken into account, the sector already accounts for a sixth of the UK economy. By 2025 it will account for about a third of the economy.

We are amazingly lucky that this sector has chosen to locate itself in the UK and given the country a considerable economic boost with relatively little official encouragement. It is critical that – though the process has some relationship to herding cats – that public policy is designed to nurture the sector and sustain its success.

CHAPTER 9

The Future of the Flat White Economy

How far will the Flat White trend develop in London, the rest of the UK and the rest of the world? Already the Flat White Economy is an amazing phenomenon that has surreptitiously changed the whole nature of the London, and to some extent the UK economy. It has driven an extraordinary rate of economic growth that has seen London growing faster than Singapore and propelled an even more extraordinary rate of job creation – a boost of 120,000 jobs mainly in the Shoreditch area. While at the same time it has created a new lifestyle based on the hip rather than on loadsamoney.

London has yet again reinvented itself. Can this continue? To look out at the future it is best to use the methodology used in earlier chapters to explain how the phenomenon came about. My understanding of how the phenomenon came into existence is that it started with technology and has been fuelled by end user demand and by the availability of extraordinary supplies of skilled labour from around the world – people who have sparked off each others' creativity. To a certain extent the phenomenon is self-fulfilling. People come to London because it is a 'happening' place. But it becomes a 'happening' place because people come to London. There is a potential instability in such self-fulfilling phenomena – if things start going wrong the laws of cumulative causation go into reverse and the decline is accelerated. But at present these laws are working in the Flat White Economy's favour and the economy is continuing to boom.

Technology

Technology relevant to the FWE is based on information and communications. Although the underlying science driving this technology have existed for more than half a century, there remains continued progress in improving the so-called 'price/performance' ratio of products driven by what is called Moore's Law.

Gordon Earle Moore was the co-founder of the Intel Corporation no less (and still its Chairman Emeritus), who in 1965 postulated that the number of transistors in a dense integrated circuit doubles roughly every two years.[1] Amazingly the prediction seems to have been proved roughly accurate even half a century on, though my friends in the industry claim that this is partly because the 'Law' is used as the benchmark for driving research activity.

Whatever the cause, this continued improvement in technological performance means that the scope for improved applications continues to grow. And the two current most highly publicised areas in IT – cloud computing and big data – are if anything even more dependent on improved technology and in turn particularly potent generators of new applications.

There has been some discussion of whether the tendency towards monopoly for parts of the IT industry such as search engines (Google), social media (Twitter and Facebook), chips (Intel), operating software (Microsoft) and hardware (Apple) will constrain growth.

We pointed out in Chapter 2 that 'supereconomies of scale' and network effects produced a tendency for 'winner take all' scenarios which is another way of describing monopoly. On the other hand, the pace of technological change in the industry creates an ever changing balance of technological advantage, which means that leadership in the sectors is always changing. The example of Nokia, which by the time it sold its handset business to Microsoft in 2013 had seen its one time market leading market share fall to 3% and its market capitalisation fall to a fifth of its peak value is instructive.[2]

As a former Chief Economist for IBM UK, my experience is that in a technological industry the position at the top of the mountain is precarious and it is surprisingly easy to lose your advantage. We used to joke that IBM stood for 'I blame Microsoft'. It is true that the modern technological leaders seem much more aggressive than the rather leisurely, slightly bureaucratic and academic IBMers. But if they are that aggressive, would they really run their businesses in such a way as to inhibit growth?

The economic analysis is inconclusive about whether a tendency to monopoly will eventually constrain the industry. But the experience of the last half century is that it hasn't happened yet – though that doesn't mean that it might do so at some future point...

End user demand

The Flat White Economy phenomenon started with the end user demand – this is driven by the UK's remarkably high proportion of online activity – the highest proportion of online retail in the world, a very high proportion of online advertising and marketing and an international competitive advantage in marketing, which means that the sector also had a significant export market of about a third of its current level of production to complement the domestic demand.

End user demand for online retail and other purchases will continue to grow. The conventional wisdom in the grocery trade is that such purchases will double over the next five years.[3] And the forecasts for the online market share from the Boston Consulting Group (Figure 2.2 in Chapter 2) show an even faster rate of growth – though I suspect these predictions are somewhat aggressive.

Growth in online retailing and increased online activity will be matched by growth of online advertising. All this means that the growth of end user demand which creates the underlying rationale for the Flat White Economy is highly likely to remain. Obviously the rate of growth will slow as the sector matures and at some point online activity will have saturated the market for online retail particularly. But that is at least ten years away and probably more. It is clear that there will be a market for the products of the Flat White Economy for many years to come.

Labour supplies and immigration

Whether Flat White activities in London continue to supply the market is much more debateable. There are two serious challenges which might mean that London could fail to compete and lose its position in the global Flat White Economy the first of which is concerns changes to immigration: the threat of a cutting off of the supply of labour.

Central to the Flat White Economy has been the remarkable flow into London of high-quality labour through immigration from all around the world but particularly from the Eurozone as Europe has remained mired in economic crisis. The European economic crisis will almost certainly continue, creating a substantial potential supply of the types of young skilled labour that have been so vital for London's growth. Youth unemployment in most parts of Continental Europe won't go away and with Europe appearing to enter a triple dip recession the potential supplies could even increase.

But will the potential migrants be allowed to come to London to boost the capital's economy? As we have seen, all the main political parties fighting the May 2015 UK general election campaigned on the basis of increased immigration controls. Meanwhile, there will be a referendum on UK withdrawal from the EU following a renegotiation of the UK's EU membership terms. Both are likely to limit incoming immigration.

Clearly the issues surrounding immigration go beyond the Flat White Economy. But one thing is clear. Starved of its labour supply of highly talented immigrants, the Flat White Economy will stagnate or worse. Politicians have a choice – they can either kowtow to public hostility to immigrants and kill the goose that lays the golden egg of economic growth or they can chose to support economic growth by resisting curbs on immigration.

Ultimately if you live in a democracy you ought to respect the decisions of the voters (right or wrong), but one of the purposes of writing this book is to show those voting starkly how great might be the costs of curbing immigration from the EU. The book is written in the hope that the public can understand what costs might be incurred if they indulge their more extreme views on immigration. And these costs appear to be very high. A complete halt to immigration from the EU would probably slow the growth of the London economy to a maximum of 2% and slow the growth of the UK economy to about 1%, both numbers roughly half the current predicted rates.

The latest estimate for UK GDP (at current market prices) for the last complete year for which we have data (2014) is £1.816 trillion.[4] My best forecast for GDP ten years hence with continued migration is for real growth of around 2% and inflation at the same rate which would give a GDP level for 2024 of £2.688 trillion. My best forecast if there is *no* immigration is for growth of around 1% and inflation of about the same pace which would give a 2024 GDP level of £2.441 trillion. In other words, the cost of restricting immigration over the next ten years would add up to an annual cost of £233 billion in ten years time. And of course the cumulative cost would be much more; though since politicians cumulate numbers to mislead and exaggerate I am deliberately not providing the cumulated number.

Now those who wish to make the anti-immigration case will quite reasonably point out that the gain to the indigenous population will be much less than this, since the extra GDP will have to be shared out amongst a larger population. This is true. But the estimated GDP is 10% higher whereas the estimated population is only 4% higher.[5] So even allowing for the fact that the GDP has to be shared over more people, GDP per capita in 2024 is forecast at £36,820 with restricted immigration and £39,128 with the current rates of immigration. In other words, if the electorate vote for very tight controls on immigration, they are voting for a 6% pay cut by 2023.

The cost of property

The other risk to the growth of the Flat White Economy is cost. Labour costs in London remain remarkably low given the cost of living as a result of the influx of labour from around the world. But the cost of accommodation is the major component of the cost of living that in turn sets the cost of labour in London. And the cost of labour cannot remain affordable forever if the cost of accommodation continues to spiral. In the short term the increased supply of accommodation in London, and particularly in the East End of London ,will permit some slackening in the pace of increase in this cost. London's average property price has risen by 7.3% in the year to Q3 2015 and the CBI/CBRE survey of policy priorities for the mayoral election places reducing housing costs as second only to improving transport as a policy priority for London's businesses.. But even these forecasts suggest that the fall will bottom out by Q4 2015 and be reversed over the subsequent two years.

To stop the cost of accommodation rising so fast that it makes London labour too expensive for the Flat White Economy, it will be necessary for housebuilding to take place on a substantial scale in London over the coming years. There are some signs that this might take place. The pace of housebuilding in London rose sharply in 2015 as Figure 9.1 shows – but has fallen back since. Yet London's housing orders are running at about six times the historic average. London Mayor Boris Johnson clearly understands the need to encourage housebuilding[7] and has been very proactive in trying to ensure that planning permissions for housebuilding developers have been granted. He has

Figure 9.1: London is undergoing a housebuilding boom for the first time since post-war reconstruction. Source ONS: New Orders in the Construction Industry.[6]

been criticised for this (including by me), but in the context of the Flat White Economy one could make the case that if he is erring, at least he is erring on the side of permitting increased economic growth.

Provided the housebuilding continues, the problem of housing accommodation will solve itself. But to keep the pace of housebuilding high, local authorities will need to be supportive in their planning decisions and in turn they will need to match the developers' spending on housebuilding with their own investment in public infrastructure. And arguably they should also complement the private sector development that is now taking place with increased public sector housing.

The other element of property costs that affects the affordability of the Flat White Economy is the cost of office accommodation. Property costs for business premises are the other main ingredient in the cost of doing business in London. When London depended mainly on financial services its business model involved limited amounts of high cost labour; expensive premises were not a major problem because they still represented a small proportion of total costs. But the Flat White Economy uses much more labour and has a much higher space requirement for each unit of value added. This means that space costs – which would have been unlikely to exceed 5% of value added in financial services – are around 30% of value added in the Flat White economy. So a shortage and resulting high price of office space would limit the growth of the Flat White Economy in a way which it never did with financial services.

The office building of 2013 and 2014 came just in time

to stabilise and slightly reduce the costs of office space in East London. Since the property crisis of 2007 the pace of London commercial development has been subdued but just as rents had started to reach dangerously high levels, a resurgence in volume started in 2013. Further growth is likely to be needed to enable the Flat White Economy sectors to grow. This in turn will again require local authorities to be liberal with their planning permissions and will also mean that the epicentre of the Flat White Economy will shift further East and South. Improved infrastructure will also be required – better road use, improved rail transport with Crossrail 2 and Crossrail 3 and improved cycle path infrastructure.

Transport

My Gresham lecture in May 2013 on how to improve transport in London[8] underlined areas of success and failure. Public transport provision in London has been a success, but its costs have been excessive. The lecture pointed out that the late Bob Crow's tube drivers earned more than Ryanair pilots, despite manifestly lower skill requirements and the combination of excessive pay and low productivity had built up an excessive cost base for all forms of public transport in London. Arguably both buses and tubes are ripe for driverless technology, particularly the latter (and where better to develop it than the home of the Flat White Economy?)

The other area where transport for London was failing was in its use of road space. As my lecture pointed out:

It is now clear that there has been a dramatic fall in

vehicle usage in London and especially in central London where the drop has been by nearly a fifth. The puzzle is that this drop in road usage has not been matched by any reduction in congestion.

Transport for London's own GPS data shows no increase in traffic speeds. And this is backed up by Cebr's European congestion study carried out for the data company INRIX. This study looked at the major cities in Europe and found by far the most congestion in London. The study showed that the average vehicle in London wasted as much as 66.1 hours each year idling in traffic jams.

Using traffic jams as a way of regulating traffic usage makes appalling economic sense. But it also makes bad environmental sense as well, since static vehicles create the very worst kinds of concentration of pollution. The Cebr study showed a cost in London of €1,896 million from the effects of stationary traffic. This does not include the cost of traffic moving much more slowly than it ought.

Transport for London themselves in the latest report on Travel in London (Report 5) are disarmingly frank about why traffic speeds have been so low when vehicle numbers have dropped – "the primary reason for the continued reductions to traffic speed, which would otherwise have been unexpected given falling traffic levels, was a substantial increase in interventions that reduced the effective capacity of the road network for general traffic." In other words, it is TfL itself who are responsible for traffic jams, according to their own research report! I bet the spin doctor would have cut that out had he bothered to read that far into the report!

A picture saves a thousand words. And the front cover

of the latest TfL report on transport demand tells the story. I think the caption says it all – even though cars are the largest single mode for personal transport in London, the cover of TfL's latest report completely ignores them. Clearly even though Boris is Mayor, the spirit of Ken lives on in TfL.

One example of the bad use of road space is bus lanes. It would not be sensible to get rid of all bus lanes, but some balance is needed in imposing them where bus usage is relatively little and where other transport users are being delayed.

The lecture also pointed out that the bad use of road space was not entirely surprising given the conflict of interest embodied in having the same transport authority controlling the use of road space as was responsible for providing public transport. Having worse traffic jams despite fewer vehicles on the road is the transport management equivalent of being unable to organise the proverbial pissup in a brewery . . . and reflects spectacular incompetence, though this is partly driven by what appears to be an ideological dislike of private motor cars.

The future and the rest of the UK

If all the four preconditions for success are met – lack of restrictions on immigration, continued housebuilding growth, more commercial development, and matching supplies of infrastructure – then the Flat White Economy will continue to be the driving force of economic growth in London and implicitly the UK.

But over time the Flat White Economy is likely to change

its nature. What currently comprises the Flat White Economy is the cutting edge of creativity in all the digital areas from high-tech to online sales to marketing. But a proportion of what is currently the cutting edge will become more mature and more business-as-usual. These activities are likely to move out of London because they will not require the creativity that is associated with London's higher cost base. Some are likely to stay within the UK because they are intimately associated with the UK environment, whether it is marketing or online retail. Others will migrate around the world to lower cost environments.

What will remain as part of the Flat White Economy in London will be the highly creative activities which require London's special ingredients. It is their scale that is uncertain and which will depend on the key issues described in the preceding paragraphs.

We showed in Chapter 5 how the Flat White Economy would be likely to spread to the rest of the UK, particularly the ten growth cities and areas that we identified. Such a development is particularly likely as the sector matures and diversifies and as a result requires different mixes of costs, creativity and disciplines. Some will be more appropriate for locations other than London.

We have seen from the analysis in Chapter 6 that if London does badly, it is not necessarily to the advantage of other parts of the UK because slower growth in London would be likely also to lead to slower growth of complementary activity in other parts of the UK. But more importantly, slower growth in London would reduce the ability to use the model of recycling the £39 billion surplus of tax over expenditure in London (nearer £50 billion if commuters are taken into account) to kick-start the economies of other regions.

The public finances model that developed particularly under Gordon Brown (both as Chancellor and Prime Minister) was to use the surplus taxation raised in London to finance public spending in other regions. This could to some extent alleviate some of the deprivation that existed outside London. But the track record of encouraging private sector activity and entrepreneurship through public spending in the UK is abysmal.

Even the left-leaning Institute for Public Policy Research was forced by the evidence to admit that: "Empirical work by Van Stel and Storey (2004)[9] looking at the relationship between the growth rate of employment in different areas of the UK and the firm birth rate in the 1980s and 1990s concludes that policies that encouraged employment growth through new firm formation appear to have had, at best, no effect on employment, and, at worst, a negative effect. This seems to be because: subsidies target the disadvantaged, who are not necessarily well-suited to running businesses; mean new firms displace existing firms in the local market; and reduce the average quality of firms being created."[10] This conclusion relates to subsidies for individual entrepreneurs but the other conclusion of the same study: "Overall, the northern regions (and the North East especially) perform poorly on the key indicators of entrepreneurship, compared with other English regions" suggests that it is the whole panoply of public policies that has failed.

Indeed, the study goes on to point out that "Alternative evidence on entrepreneurship is available from the Household Survey of Entrepreneurship conducted by the

Small Business Service. A regional breakdown of attitudes to entrepreneurship in 2005 suggests that the North East and North West are slightly less entrepreneurial than other English regions. Evidence from the Global Entrepreneurship Monitor survey for 2006 suggests that the northern regions have lower proportions of their working-age population engaged in business start-ups than the UK average".

My instinctive assessment of public spending is that it is a very weak rod to encourage business activity, partly because of the reasons set out by the left-wing think tank but also because the spending tends to have political rather than economic objectives (especially in the UK – bizarrely, it seems to work better in some other countries where a degree of corruption seems to generate a higher degree of harmony between political and economic objectives!)

In Chapter 6, I proposed that the model of using the tax surpluses in London to subsidise public spending in the rest of the country should be replaced by a new policy of using these surpluses to pay for lower taxes on businesses outside London. These could be in the form of a lower rate of corporation tax, but might equally be a reduction in the employers' NI contributions or lower business rates. Clearly the last of these would be most easy to target geographically, while lower NI contributions would target better employment. Whichever approach is used, the scope is considerable – the £50 billion surplus in London and the South East would pay for a subsidy of £2,500 per private sector employee outside the South of England (London, South East, East and South West). As someone

who is not only an economist but also a successful entrepreneur I can guarantee that a subsidy on this scale would
seriously boost entrepreneurship.

Simulations in the US suggest that when applied to the
UK, subsidies on this scale could potentially create up to a
million new jobs[11]. So if there is a will to make the rest of
the UK as successful as London it can be done.

Other countries

Silicon Valley, Boston and in a different way Bangalore
are world renowned for their digital economies. But we
have shown that there is a range of cities in the US with
growing digital economies. The latest group to start to
present themselves are the rust belt cities of the Northern
mid-west. Detroit might take some time to reinvigorate
itself, with unpaid city workers and feral dogs roaming
the streets but Chicago, Minneapolis and even Buffalo,
famously written off by economist Edward Glaeser, look
like candidates for success. And of course Silicon Valley is
not the only part of the West Coast to be host to a digital
economy – both Seattle and Portland Oregon are already
successfully driving Flat White Economies of their own.
In Canada, Vancouver, Montreal and Toronto are strong
candidates; in Australia, Sydney especially.

In Europe the highest level of government promotion
of ICT is in France. But with the French in serious budget deficit problems and with spectacularly high tax rates
one suspects that achieving success here is the equivalent of pushing water uphill. I would expect Stockholm,
Copenhagen, Barcelona and Berlin to be more likely to
mimic London's success than Paris.

London's advantages of access to large amounts of migrant labour and the widespread use of the English language are shared with the major Scandinavian cities. This is why Stockholm particularly and also Copenhagen are now emerging as significant Flat White Economies in their own right. On a per capita basis Stockholm is second only to Silicon Valley in startup numbers with Spotify as the star example. The Nordic economies count for 2% of world GDP but 10% of new companies sold for more than a billion dollars[12].

In Asia, the more conformist educational systems and reduced willingness to challenge conventional wisdom will hold the sector back. In purely statistical terms, Asia ought to be much more likely to develop a Flat White Economy than either the US or Europe. The technology is there, end user demand is advanced and educational levels in a formal sense are high, while labour supplies are available cheaply. But a conformism in thinking patterns and – in most places – limited amounts of cross fertilisation through migration will hold many Asian cities back. In addition they will be hampered by lack of protection of intellectual property in many parts of the region. My best bets are Calcutta, Mumbai, Singapore, Kuala Lumpur, Hong Kong, Beijing, Tokyo and Seoul (Gangnam style). And of course Bangalore which is already there …

But it is important when looking to the future to realise that chance plays a major role in deciding which centre does best. London had the good fortune to have a mass supply of labour from the economic collapse in Europe becoming available, just at the same time as demand for online retailing and marketing were taking off in the UK and cheap warehouse potential office space in Shoreditch had finally started to be converted.

I am sure that more than one of the cities I have mentioned above will benefit from similar good luck in some way. Indeed, I would be more than delighted in a decade's time to read a book about the Flat White Economy in Seoul, Calcutta, Singapore, Kuala Lumpur, Sydney, Barcelona or Berlin. Or somewhere like Ulan Bator, Samarkand, Addis Ababa or Ouagadougou that is currently completely off the radar!

NOTES

FORWORD

1. It is indicative of the volatility of technology shares that by 1 Feb 2016, Alphabet, the new parent company of Google, had overtaken Apple to become the world's largest company with a market capitalisation of $550 billion compared with Apple's $530 billion.

PROLOGUE

1. I have dropped the company registration data showing 32,000 startups in the two years to March 2014 from this edition because I have been advised that the data is unreliable. The figures in the earlier edition incorporate a surprising number of registrations at a single address. However, the overall dynamism of the area is backed up by other data about corporates including PAYE data. I am indebted to Professor Ian Gordon of the LSE for bringing this point to my attention.
2. http://www.techworld.com/news/startups/startups-still-piling-into-offices-near-silicon-roundabout-3625434/ The average number for the rest of London is 58 tech firms per km
3. www.centreforcities.org/assets/files/2014/Cities_Outlook_2014.pdf
4   www.theguardian.com/world/2014/jun/18/britain-old-declining-empire-official-chinese-newspaper
5. Census and Statistics Department, the Government Office of the Hong Kong Special Administrative Region

CHAPTER ONE

1. Property is always going to be important in areas where, for either geographical or planning reasons, there is a shortage of land. Many of the key companies in Hong Kong and Singapore are property-based for this

reason. The property industry is a critical one for London today – by and large London has been well-served by its developers, but with property so intrinsic to economic cycles, it has been a rollercoaster ride.

2. Cebr has collated a series of statistics for City jobs which includes jobs in wholesale financial services elsewhere in London but excludes jobs in the City (such as retail) which are not directly connected with financial services.

3. An estimation based on a survey of how the rich spent their money: 'The Rich – Are they Different?', Douglas McWilliams and Mark Pagnell, *Prospect*, October 1995.

4. Statistics on house prices from www.lloydsbankinggroup.com/media/economic-insight/regional-house-prices.

5. London's Economic Outlook: Autumn 2015 The GLA's medium-term planning projections GLA Economics

6. For regional accounts, the Gross Domestic Product (GDP) for a region like London is conventionally measured as Gross Value Added (GVA). The concepts are essentially the same but the valuation is different. GVA is measured at producers' basic prices, since it is normally compiled from the outputs of industries. GDP is measured at market prices. If you add taxes and subtract subsidies from the GVA data you get the GDP for the region.

7 www.ons.gov.uk/ons/rel/bus-register/business-register-employment-survey/2013-provisional/info-jobs-growth-by-region-and-industry.html

8 See: http://www.cityam.com/226279/embargo-cebr-london-house-prices-are-costing-londons-economy-1bn-each-year

9 http://www.theguardian.com/money/2015/jan/25/london-tenants-forced-to-share-rooms

10 http://www.handyshippingguide.com/shipping-news/road-haulage-lobby-informs-government-and-welcomes-infrastructure-changes-affecting-freight-vehicles_6710

CHAPTER TWO

1. 'The Economics of Information', George J Stigler, *Chicago University Journal of Political Economy*, June 1961, pp213–225.

2. Originally published as 'Metcalfe's Law and Legacy', *Forbes ASAP*, September 1993, and developed in *Telecosm*, George Gilder, Simon & Schuster, 1996.

3. I first presented these arguments with some accompanying detail to the IBM Computer Users Association in May 1989 – they have changed remarkably little in the past 25 years!

4. 'The Economic Impact of Cloud Computing: a Cebr report for EMC²', Cebr, December 2010.

5. 'The Value of Data Equity: A Cebr report for SAS', Cebr, April 2012.

6. 'Britons are the biggest online shoppers in the developed world', James Hall, *Daily Telegraph*, 1 Feb 2012.

7. www.internetretailing.net/2014/03/uk-retailers-expected-to-make-online-sales-of-45bn-this-year-study/

8. 'Advertising Pays', Deloitte, Jan 2013:www.adassoc.org.uk/pdfs/advass_advertising_pays_report.pdf

9. 'The Business of Evidence: A report prepared for the Market Research Society', Price Waterhouse Coopers, October 2012:www.mrs.org.uk/pdf/The_Business_of_Evidence_Final_08102012.pdf

10. www.cityoflondon.gov.uk/business/economic-research-and-information/statistics/Documents/FS-PBS-%20April2013.pdf Study by Oxford Economics Table 2

11. 'Creativity in Advertising: When It Works and When It Doesn't', W Reinartz & P Saffert, *Harvard Business Review*, June 2013: hbr.org/2013/06/creativity-in-advertising-when-it-works-and-when-it-doesnt

12. Tim Wood, *Stop Press*, 26 July 2013: www.stoppress.co.nz/blog/2013/07/how-effective-creativity-advertising

13. 'Creativity in Advertising', William M O'Barr, *Advertising & Society Review*, Vol 11, no. 4, 2011: muse.jhu.edu/journals/advertising_and_society_review/v011/11.4.o-barr01

14. Edwidge Danticat (in interview), *The Atlantic*, 27 August 2013

15. *Diasporas: Concepts, Intersections, Identities*, Kim Knott & Seán McLoughlin (eds), Zed Books, 2010.

16. 'Skilled immigration and strategically important skills in the UK economy' (final report to the Migration Advisory Committee), A George, M Lalani, G Mason, H Rolfe, C Rosazza Bondibene, 2012: www.niesr.ac.uk/pdf/290212_151752.pdf

17. 'Transnational communities and the evolution of global production networks: the cases of Taiwan, China and India', Annalee Saxenian, *Industry and Innovation* (a special issue on Global Production Networks), 2002.

18. 'The Impact of Recent Immigration on the London Economy', London School of Economics, July 2007.

19. 2011 Census (workplace population analysis), Office for National Statistics, May 2014: www.ons.gov.uk/ons/dcp171766_364058.pdf

20. www.bbc.co.uk/news/uk-england-25879675

21. 'Internal Migration by Local Authorities in England and Wales, Year Ending June 2012', Office for National Statistics, June 2013: www.ons.gov.uk/ons/dcp171778_315652.pdf

22. 'Simply the Best? Skilled migrants and the UK's knowledge economy', L Hopkins & C Levy, The Big Innovation Centre, June 2012.
23. Under the UK's national qualifications framework, Level 4 is equivalent to a Higher National Certificate – see www.gov.uk/what-different-qualification-levels-mean.
24. 2011 Census (workplace population analysis), Office for National Statistics, May 2014: www.ons.gov.uk/ons/dcp171766_364058.pdf
25. travel.wikinut.com/The-Cultural-Diversity-of-London/y6e37vl3/
26. 'Proof that Diversity drives Innovation', Donald Fan, Diversity Inc: www.diversityinc.com/diversity-management/proof-that-diversity-drives-innovation
27. 'Information and Structural Holes', Ronald S Burt, *Industrial and Corporate Change*, Vol 17 No 5, pp953–969, August 2008.
28. 'Britain's Cosmopolitan Capital: The World comes to London', *The Economist*, 7 August 2003.
29. neighbourhood.statistics.gov.uk

CHAPTER THREE

1. Data from Companies House 'Incorporated Companies in the United Kingdom September 2015'
2. Source: UK Business: Activity, Size and Location 2015, ONS. Note that the number of businesses recorded has risen during this period by an estimated 31,000 due to improved measurement. On yet another measure the total number of businesses in the UK is estimated to have risen from 3.47 million in 2000 to 5.49 million in 2015 (source House of Commons Library Briefing Paper Number 06152, 20 October 2015 by Chris Rhodes).
3. Annual Business Survey: UK Non Financial Business Economy release 23/07/2015
4. http://www.techcityuk.com/wp-content/uploads/2015/02/Tech%20Nation%202015.pdf
5. https://en.wikipedia.org/wiki/East_London_Tech_City
6. http://www.techcityuk.com/investors/
7. www.gov.uk/government/speeches/east-end-tech-city-speech
8. www.gov.uk/government/speeches/cebit-2014-david-camerons-speech
9.. The most embarrassing error in the initial version of this book was my use of an old IBM definition of 'fintech' to suggest that it meant the business of financing technology! Mea culpa! Fintech is the application of technology to finance.
10. 'Characteristics of SMEs and Social Enterprises around Tech City',

Experian (prepared for the City of London Corporation), November 2012.

11. 'The City of London's Supply Chain: A scoping study for the Analysis of the Relationship between the City and its Fringes', Cebr (prepared for the Corporation of London), p9.
12. p7 *Ibid.*

CHAPTER FOUR

1. www.theguardian.com/travel/2014/jun/30/london-best-coffee-shops-readers-tips
2. www.standard.co.uk/news/coffee-shops-just-keep-on-buzzing-7303853.html
3. www.hospitalityandcateringnews.com/2013/01/uk-coffee-market-grows-and-to-grow-strongly
4. www.theguardian.com/lifeandstyle/2012/dec/22/pubs-coffee-shops
5. www.telegraph.co.uk/finance/newsbysector/retailandconsumer/leisure/11084328/Why-coffee-shops-are-replacing-pubs-in-Britain.html
6. www.hospitalityandcateringnews.com/2013/01/uk-coffee-market-grows-and-to-grow-strongly
7. www.ft.com/cms/s/2/11976c50-8ae8-11df-bead 00144feab49a.html#axzz3GsIB94bG
8. www.telegraph.co.uk/finance/newsbysector/retailandconsumer/leisure/11084328/Why-coffee-shops-are-replacing-pubs-in-Britain.html
9. Data from the British Beer and Pubs Association: www.beerandpub.com/statistics
10. This analysis was carried out by Cebr staff.
11. twowheelsgood-fourwheelsbad.blogspot.co.uk/2013/06/cycling-is-now-dominant-mode-of-travel.html
12. data.london.gov.uk/datastore/package/cycle-flows-tfl-road-network
13. www.ctc.org.uk/resources/ctc-cycling-statistics
14. www.bikebiz.com/news/read/cycling-isn-t-mainstream-enough-yet/016851
15. London has 29 inches of rainfall annually, Amsterdam has 31 www.skyscrapercity.com/showthread.php?t=349393
16. amsterdamize.com/2011/11/21/bicycle-cultures-are-man-made
17. www.newscientist.com/article/dn24636-despite-the-deaths-cycling-in-london-is-getting-safer.html#.VETGl_nF-So
18. www.businessinsider.com/peter-thiel-ama-2014–9#ixzz3GflCaosA
19. www.esquire.co.uk/culture/features/5720/the-silicon-roundabout
20. The article attributes the phrase Silicon Roundabout to a tweet from software developer Matt Biddulph working out of Moo.com in July 2007.

21. www.theguardian.com/technology/2013/nov/01/google-new-london-headquarters
22. Quote from *The Guardian* article above.
23. *Ibid.*
24. www.neighbourhood.statistics.gov.uk/HTMLDocs/dvc126/
25. s3.amazonaws.com/bbpa-prod/attachments/documents/resources/22617/original/Oxford%20Economics%20for%20the%20BBPA%20Regional%20Impacts%20Jan%202014.pdf?1391684386
26. www.fabians.org.uk/the-tory-feelgood-factor/
27. Calculations by Cebr from data from the Champagne Bureau.

CHAPTER FIVE

1. www.digitalcommunities.com/survey/cities/?year=2013
2. www.forbes.com/sites/joelkotkin/2012/05/17/the-best-cities-for-tech-jobs
3. www.technologyreview.com/news/517626/infographic-the-worlds-technology-hubs
4. www.technologyreview.com/news/516501/in-innovation-quest-regions-seek-critical-mass
5. It is important not to be too absolutist in the debate about government assistance. Even the Flat White Economy near Old Street Roundabout, which was not centrally planned and where the development has been essentially private sector led, has benefitted from public investment in transport hubs and support from local authorities, including especially the City of London, to regenerate the so-called City Fringes.
6. www.epps.fr/en/a-global-cluster/innovation-growth-employment
7. www.grand-paris.jll.fr/id/155
8. www.polytechnique.edu/en/paris-saclay
9. www.nature.com/news/2010/101020/full/467897a.html
10. www.themoscowtimes.com/business/article/skolkovo-foundation-to-get-15bln-in-2013–2020/483982.html
11. www.themoscowtimes.com/business/article/skolkovo-foundation-to-get-15bln-in-2013–2020/483982.html
12. www.slate.com/articles/technology/the_next_silicon_valley/2013/12/russia_s_innovation_city_skolkovo_plagued_by_doubts_but_it_continues_to.html
13. www.slate.com/articles/technology/the_next_silicon_valley/2013/12/russia_s_innovation_city_skolkovo_plagued_by_doubts_but_it_continues_to.2.html
14. data.worldbank.org/indicator/GB.XPD.RSDV.GD.ZS?order=wbapi_data_

value_2012+wbapi_data_value+wbapi_data_value-last&sort=desc

15. www.israel21c.org/technology/innovation/made-in-israel-the-top-64-innovations-developed-in-israel

16. www.forbes.com/sites/realspin/2013/11/07/what-are-the-secrets-behind-israels-growing-innovative-edge

17. www.ft.com/cms/s/0/166799a0-fdda-11e1-9901-00144feabdc0.html#axzz3IfWNfKmd

18. immigrationlaw.goop.co.il/Web/?PageType=0&ItemID=43522

19. The Indian name for the City is Bengaluru. While most now refer to the former Bombay as Mumbai, and to the former Madras as Chennai, the convention appears to be to continue to call Bangalore by its Westernised name and I have done so here. This presumably reflects the extent of the use of the English language and the international nature of the city.

20. www.bbc.co.uk/news/technology-23931499

21. blog.lewispr.com/2014/01/top-tech-capitals-in-the-world-bangalore-the-it-capital-of-india.html

22. profit.ndtv.com/news/industries/article-bangalore-leads-growth-in-creating-new-jobs-369463

23. timesofindia.indiatimes.com/tech/tech-news/Bangalore-among-top-8-tech-clusters/articleshow/21544103.cms

24. www.theguardian.com/global-development/poverty-matters/2013/apr/17/bengaluru-rues-rapid-growth-india-it. See also econ.worldbank.org

25. In my Gresham lecture 'How Does Globalisation affect Inequality Globally?' (www.gresham.ac.uk/lectures-and-events/how-does-globalisation-affect-inequality-globally), I pointed out how globalisation has been associated with a reduction in global poverty to twice the extent targeted in the Millennium Development Goals (established in 2000) – i.e. halved, which was the original goal, and then halved again within the time period from 1990 to 2015. I remain amazed that this achievement is hardly ever mentioned in the left-wing press, which one might have expected to have some interest in world poverty.

26. www.forbes.com/sites/ruima/2014/10/20/one-billion-chinese-entrepreneurs

27. *Ibid*

28. www.cyberport.hk/en

29. www.forbes.com/sites/china/2010/05/11/why-hong-kong-is-chinas-new-tech-hub

30. Authorised Depositary Receipts – the equivalent of shares for suitable foreign companies in the US.

31. www.yoursingapore.com/content/mice/en/why-singapore/key-industry-sectors/media-and-digital-content.html

32. www.globalinnovationindex.org/content.aspx?page=data-analysis

33. www.yoursingapore.com/content/mice/en/why-singapore/key-industry-sectors/innovation.html
34. Wikipedia calls Bangalore "The Pub Capital of India" en.wikipedia.org/wiki/Bangalore
35. www.theguardian.com/small-business-network/2014/feb/24/birmingham-new-technology-businesses
36. 'The economic contribution of the media sector in Glasgow': Report for the Glasgow Chambers of Commerce April 2014
37. www.ft.com/cms/s/2/ad9ab0a2-9e1e-11e2-bea1-00144feabdc0.html#axzz2Umx03300
38. features.techworld.com/sme/3589108/edinburgh-scale-up-fanduel-rejected-by-80-investors-on-way-to-becoming-1bn-firm/
39. www.ft.com/cms/s/0/58d74174-7381-11e2-9e92-00144feabdc0.html#axzz3NC4cAj7g
40. www.cebr.com/reports/uk-local-innovation-index

CHAPTER SIX

1. *Triumph of the City: How Our Greatest Invention Makes Us Richer, Smarter, Greener, Healthier, and Happier*, Edward Glaeser, 2011, Penguin.
2. The details are contained in the 'Report From The Joint Select Committee Of The House Of Lords And The House Of Commons On The Port Of London Bill': Joint Select Committee, Houses of Parliament.
3. Personal correspondence.
4. I have used England rather than the UK as a reference point here for consistency, because of past changes in the status of Ireland within the UK, and possible future changes in the status of Scotland within the UK over the period.
5. *Four Thousand Years of Urban Growth: An Historical Census*, Tertius Chandler, 1987, St. David's University Press, Lewiston, NY.
6. 'Population and Employment Projections to Support the London Infrastructure Plan 2050', GLA Economics (update) November 2013 (www.london.gov.uk/sites/default/files/Population%20and%20employment%20projections%20to%20support%20the%20London%20Infrastructure%20Plan%202050.pdf).
7. 'Regional GDP in Britain, 1871–1911: Some Estimates' Nicholas F. R. Crafts ,Working Paper No. 03/04 ,Department of Economic History London School of Economics, March 2004.
8. www.zerohedge.com/news/2014–02–20/uks-2-tier-economy-london-and-everyone-else
9. 'The London Problem', Danny Dorling, *New Statesman* 29 August – 04 September 2014, pp27-31.

10. This concept is described in most economics textbooks. See for example: www.princeton.edu/~achaney/tmve/wiki100k/docs/Pareto_efficiency.html

11. See for example the website: Inequality.org.

12. www.economist.com/news/britain/21637420-green-party-growing-force-british-politics-if-only-it-was-more-world-green (3 January 2015)

13. www.economist.com/news/britain/21637420-green-party-growing-force-british-politics-if-only-it-was-more-world-green (3 January 2015)

14. www.independent.co.uk/news/uk/politics/green-party-leader-i-didnt-say-being-on-benefits-in-britain-was-worse-than-being-poor-in-india-9950573.html

15 See for example 'Measuring Government in the Twenty First Century' by Livio de Matteo, 2013 from the Fraser Institute in Canada. This postulates a quadratic relationship between the share of public spending and economic growth with an optimal rate of about 20% of GDP. Until that point, increases in public spending are mainly infrastructure and education – investments that boost growth. Beyond that point a much higher proportion of public spending is consumption and welfare, some of which directly inhibits growth. Meanwhile, increases in taxation above 20% of GDP or so start to reduce the returns on investment and entrepreneurship and hence discourage wealth creation (www.fraserinstitute.org/uploadedFiles/fraser-ca/Content/research-news/research/publications/measuring-government-in-the-21st-century.pdf)

16. The most extreme example of this of which I am aware is Hong Kong, where arguably most of the wealth creation in recent years has not benefitted the bulk of the population because of scarce land and rising (quasi-monopolised) landholdings which has meant that the benefits of higher growth have been eaten away in higher rents. The ratio of property prices to average earnings in Hong Kong is roughly twice as high as in London.

17. http://www.dazeddigital.com/artsandculture/article/24336/1/this-map-marks-all-of-londons-anti-gentrification-campaigns

18. I have to declare an interest – I wrote this report!

19. This report was prepared for the Greater London Authority (legacy.london.gov.uk/mayor/economic_unit/docs/growing_together_report.pdf).

20. This report was also prepared by Oxford Economic Forecasting for the Corporation of London (A copy can be found at www.bipsolutions.com/docstore/pdf/7493.pdf).

21. This report was prepared by Leticia Veruete-McKay in the GLA Economics Department (www.london.gov.uk/sites/default/files/current_issues_note_5.pdf).

22. See for example The Dangers of Decentralisation Rémy Prud'homme,

The World Bank Research Observer, vol. 10, no. 2 August 1995. pp201-20. (www-wds.worldbank.org/external/default/WDSContentServer/WDSP/IB/2013/05/13/000333037_20130513130514/Rendered/PDF/770740JRN0WBRO0Box0377291B00PUBLIC0.pdf).

23. This report was prepared for the South East of England Regional Assembly (May 2005) by Cebr staff although I was not very much involved personally; however its methodologies were based on those that I had pioneered

24. 'London's Finances and Revenues', Corporation of London November 2014. The report was mainly written by my colleagues at Cebr, specially Charles Davis and Rob Harbron for the City of London Corporation, although I was consulted on methodology and also reviewed the conclusions (www.cityoflondon.gov.uk/business/economic-research-and-information/research-publications/Documents/Research-2014/londons%20finance%20and%20revenues.pdf).

25. Source: Cebr calculations.

26. *Ibid.*

27. Source: Author's calculations based on Cebr data in previous two charts.

28. 'Business Statistics', House of Commons Library Standard Note SN/EP/6152, 14 July 2014, updated 28 November 2014 by Chris Rhodes (www.parliament.uk/briefing-papers/sn06152.pdf).

CHAPTER SEVEN

1. *The Peopling of London: Fifteen Thousand Years of Settlement from Overseas Museum of London,* 1996, Sara Selwood, Bill Schwartz and Nick Merriman, (The Poitevins were settlers from the French town of Poitiers).

2. www.lse.ac.uk/geographyandenvironment/research/london/pdf/theimpactofrecentimmigrationonthelondoneconomy.pdf

3. The secondary source is *The Peopling of London*, Roy Porter. See above.

4. *London: A Social History*, Penguin, 1994.

5. www.bbc.co.uk/news/uk-politics-29684585

6. blogs.spectator.co.uk/coffeehouse/2014/11/david-camerons-immigration-speech-full-text

7. www.express.co.uk/news/politics/521904/Ed-Miliband-will-vow-to-crack-down-on-immigration

8. http://www.libdems.org.uk/immigration

9. http://www.ukip.org/ukip_launches_immigration_policy

10. I must declare an interest here – from 1974 to 1985 I worked as an economist for the CBI and from 1989 to 1993 I was the Chief Economic Adviser (with the title of Deputy Director General) to the organisation.

11. www.ft.com/cms/s/0/cbc29f68-7780-11e3-807e-00144feabdc0.

html#axzz3IOHaYqKJ *Financial Times* 7 January 2014

12. www.gov.uk/government/uploads/system/uploads/attachment_data/file/287287/occ109.pdf

13. 'Impacts of migration on UK native employment: An analytical review of the evidence' Ciaran Devlin and Olivia Bolt, Department for Business, Innovation and Skills Dhiren Patel, David Harding and Ishtiaq Hussain, Home Office, March 2014.

14. This is one of the most prestigious of the regular academic economics publications. The article is 'The Fiscal Effects of Immigration to the UK' by Christaan Dustmann and Tommaso Frattini, *The Economic Journal*, vol. 124, issue 580, ppF593–F643, John Wiley, November 2014.

15. This is simply the accumulated budget deficits over the period adjusted for the contribution of the migrants.

16. 'Migration and productivity: employers' practices, public attitudes and statistical evidence' Heather Rolfe, Cinzia Rienzo, Mumtaz Lalani and Jonathan Portes, National Institute for Economic and Social Research November 2013.

17. niesr.ac.uk/sites/default/files/publications/Migration%20productivity%20final.pdf

18. www.kpmg.com/UK/en/IssuesAndInsights/ArticlesPublications/NewsReleases/Pages/Half-of-businesses-in-London-suffering-from-skills-shortage-warns-CBI-KPMG.aspx

19. http://news.cbi.org.uk/news/capital-s-firms-to-next-mayor-transport-and-housing-must-improve-cbi-cbre/cbi-cbre-london-business-survey/

20. At various points in my life I was directly in charge of the CBI's business surveys including a five-year period in the late 1970s and early 1980s. In those days, when such research was well resourced, we used to conduct surveys on answering practices and it was clear that the majority of respondents when they claimed that output was limited by lack of skills had specific examples in mind. For more detail see *Twenty Five Years of Ups and Downs* (1988), edited by the author, CBI, London.

21. www.gresham.ac.uk/lectures-and-events/was-karl-marx-always-wrong

22. *British Capitalism, Workers and the Profit Squeeze,* Andrew Glyn and Bob Sutcliffe, 1972, Penguin.

23. I remember Sir Ralph Bateman, CBI President and Chairman of Turner and Newall, telling the then Chancellor of the Exchequer Denis Healey, in October 1974 with a straight face that: "Inflation on the current scale could see the end of capitalism as we know it. But inflation without profits was even worse!"

24. In June 1980 I gave a paper to the Georgetown University Centre for Strategic and International Studies seminar on the Thatcher experiment. Essentially this was the Reagan team hoping to be elected that Autumn trying to see what had worked from the beginnings of the Thatcher

'experiment' as they saw it. My conclusion was that because the payoff from Thatcherite policies would take a number of years, the critical factor was whether Mrs Thatcher could win a second election to avoid the policies being reversed before the benefits had emerged. Trade Union leader Frank Chapple, also attending the seminar, told me I had got it right. But only a combination of Michael Foot as Labour leader, the Falklands War and the split in the Labour party to form the SDP made Mrs Thatchers' 1983 victory inevitable.

25. www.oecd.org/g20/topics/employment-and-social-policy/G20-labour-markets-outlook-key-challenges-and-policy-responses.pdf

26. The World at Work: Jobs, Pay and Skills for 3.5 Billion People, R. Dobbs, A. Madgavkar, D. Barton, E. Labaye, J. Manyika, C. Roxburgh, S. Lund, S. Madhav, McKinsey Global Institute, June 2012.

27. www.oecd.org/statistics/

28.. www.oecd.org/g20/topics/employment-and-social-policy/G20-labour-markets-outlook-key-challenges-and-policy-responses.pdf

29. 'The Global decline of the Labor Share' Loukas Karabarbounis and Brent Neiman, Working Paper 19136, NBER, June 2013: www.nber.org/papers/w19136.

30. *The Economic Consequences of the Peace* John Maynard Keynes, 1919, London. (Numerous editions are available in print, eBook and PDF. This passage is taken from 'Chapter II: Europe Before the War: III, The Psychology of Society', and can be found at www.econlib.org/library/YPDBooks/Keynes/kynsCP2.html

31. *Das Capital* (Vol I, Ch 24, Sect 3), Karl Marx, 1867 (See www.marxists.org/archive/marx/works/download/pdf/Capital-Volume-I.pdf).

32. highgrowthsmallbusiness.co.uk

33. 'Cyclicality of SME Finance: Literature Survey, Data Analysis and Econometric Analysis' EIM Business Policy and Research, March 2009 (ec.europa.eu/enterprise/newsroom/cf/itemdetail.cfm?item_id=3157). This EU report on funding for SMEs draws attention to the complementarity of internal and external funding.

CHAPTER EIGHT

1. It is one of the ironies of British politics that the so-called 'egalitarian' tax policies of both Labour and the Coalition have left Britain with a tax system which has taken the poor out of tax and placed very high taxes on the rich. The result is a tax system that now actually requires a high level of inequality to raise significant tax revenues. Surely this is one of the more extreme examples of the law of unintended consequences!

2. The latest data is available from the following website www.gov.uk/government/statistics/input-output-supply-and-use-tables-2014-edition

3. "We need not look beyond the Government's creation of a "candy-floss" economy in this country". *Hansard* 6 February 1964. It is a worthwhile exercise to read *Hansard* from the 1960s. What comes across is how seriously the politicians of the time took their debates and how rigorously they focused on economic policy and its implications. Today Parliamentary debate (at least when in the public eye) is essentially point scoring against the other side and what passes for debate is in reality an overdose of cheap jibes. But you only have to go back to the 1960s to discover that there really was a Golden Age of Parliamentary debate.

4. 'Understanding National Accounts', François Lequiller and Derek Blades, 2016, Paris (www.oecd.org/std/na/38451313.pdf).

5. www.bankofengland.co.uk/statistics/Pages/iadb/notesiadb/capexp.aspx

6. The productivity paradox of information technology', Erik Brynjolfsson, *Communications of the ACM* 36 (12), 1993, pp66–77.

7. 'We'd better watch out', Robert Solow, *New York Times Book Review*, 12 July, 1987, p.36.

8 scholar.google.co.uk/ scholar?start=20&q=information+technology+as+an+enabler&hl=en&as_sdt=0,5&as_vis=1

9. 'Sources of Economic Growth, Trade and Investment Analytical Papers No 6 of 18,BIS/DFID, 2011, London.

10. www.gov.uk/government/uploads/system/uploads/attachment_data/file/32468/11-723-sources-of-economic-growth.pdf

11. 10 November 2014.

12. www.itpro.co.uk/mobile/23478/o2-ceo-how-the-digital-revolution-is-driving-the-uk-economy#ixzz3K5a452gr

13. GES Group on growth, 'Innovation', November 2014: www.gov.uk/government/uploads/system/uploads/attachment_data/file/370187/bis-14-1169-government-economic-service-group-on-growth-innovation.pdf

14. Actually the chart is too busy even to make much sense in the original report …

15. *Diffusion of Innovations*, Everett M. Rogers, 1962, Free Press, Glencoe.

16. www.oecd-ilibrary.org/science-and-technology/frascati-manual-2002_9789264199040-en

17. www.gov.uk/government/uploads/system/uploads/attachment_data/file/293635/bis-14-p188-innovation-report-2014-revised.pdf. Details from Section 3.1.

18. 'Innovation Report: innovation, Research and Growth', BIS, March 2014, p.3 (as above).

19. 'Creative Britain The current and future contribution of the creative industries to the UK economy', Report for Falmouth University, Cebr,

November 2014 (www.falmouth.ac.uk/sites/default/files/download/fal-mouth_cebr_report_final.pdf).

CHAPTER NINE

1. 'Cramming more components onto integrated circuits', Gordon E. Moore, *Electronics Magazine*, 1965 p.4.
2. There is a neat summary of this in *The New Yorker*, 3 September 2013 (www.newyorker.com/business/currency/where-nokia-went-wrong).
3. www.intelligentpositioning.com/blog/2013/09/uk-online-grocery-sales-up-set-to-double-in-5-years
4. Office for National Statistics (www.ons.gov.uk/ons/rel/naa2/second-estimate-of-gdp/q3-2014/tsd-second-estimate-of-gdp--q3-2014.html).
5. I have taken the population assumptions from the Migration Observatory in Oxford (www.migrationobservatory.ox.ac.uk/briefings/impact-migration-uk-population-growth).
6. www.ons.gov.uk/ons/taxonomy/index.html?nscl=New+Orders+in+the+Construction+Industry#tab-data-tables
7. As I have found to my cost when I was arguing on behalf of Islington Council that the planning gain to a particular developer from being allowed to make a major redevelopment without adequate provision of social housing was excessive, causing my first ever defeat as an expert in a legal or planning case!
8. www.gresham.ac.uk/lectures-and-events/sorting-out-transport-in-london
9. 'The link between firm births and job creation: Is there a Upas tree effect?' A.J. Van Stel and D.J. Storey, *Regional Studies* 38(8), pp893-909.
10. 'Entrepreneurship and Innovation in the North' Paper 3, Northern Economic Agenda Project, Michael Johnson and Howard Reed, January 2008. (Institute for Public Policy Research www.ippr.org/assets/media/images/media/files/publication/2011/05/entrepreneurship_and_innovation_1619.pdf).
11. See for example research.upjohn.org/cgi/viewcontent.cgi?article=1021&context=confpapers
12. http://www.creandum.com/nordic-tech-is-on-fire-almost-10-of-all-busd-exits-last-10-years/

INDEX